D1234747

Mathematical Cosmology and Extragalactic Astronomy

Pure and Applied Mathematics

A Series of Monographs and Textbooks

RECENT TITLES

XIA DAO-XING. Measure and Integration Theory of Infinite-Dimensional Spaces: Abstract Harmonic Analysis

RONALD G. DOUGLAS. Banach Algebra Techniques in Operator Theory

WILLARD MILLER, JR. Symmetry Groups and Their Applications

ARTHUR A. SAGLE AND RALPH E. WALDE. Introduction to Lie Groups and Lie Algebras

T. BENNY RUSHING. Topological Embeddings

JAMES W. VICK. Homology Theory: An Introduction to Algebraic Topology

E. R. KOLCHIN. Differential Algebra and Algebraic Groups

GERALD J. JANUSZ. Algebraic Number Fields

A. S. B. HOLLAND. Introduction to the Theory of Entire Functions

WAYNE ROBERTS AND DALE VARBERG. Convex Functions

A. M. OSTROWSKI. Solution of Equations in Euclidean and Banach Spaces, Third Edition of Solution of Equations and Systems of Equations

H. M. EDWARDS. Riemann's Zeta Function

SAMUEL EILENBERG. Automata, Languages, and Machines: Volume A. *In preparation:* Volume B

MORRIS HIRSCH AND STEPHEN SMALE. Differential Equations, Dynamical Systems, and Linear Algebra

WILHELM MAGNUS. Noneuclidean Tesselations and Their Groups

J. DIEUDONNÉ. Treatise on Analysis, Volume IV

FRANÇOIS TREVES. Basic Linear Partial Differential Equations

WILLIAM M. BOOTHBY. An Introduction to Differentiable Manifolds and Riemannian Geometry

BRAYTON GRAY. Homotopy Theory: An Introduction to Algebraic Topology

ROBERT A. ADAMS. Sobolev Spaces

JOHN J. BENEDETTO. Spectral Synthesis

D. V. WIDDER. The Heat Equation

IRVING EZRA SEGAL. Mathematical Cosmology and Extragalactic Astronomy

In preparation

WERNER GREUB, STEPHEN HALPERIN, AND RAY VANSTONE. Connections, Curvature, and Cohomology: Volume III, Cohomology of Principal Bundles and Homogeneous Spaces

J. DIEUDONNÉ. Treatise on Analysis, Volume II, enlarged and corrected printing

I. MARTIN ISAACS. Character Theory of Finite Groups

Mathematical Cosmology and Extragalactic Astronomy

IRVING EZRA SEGAL
Department of Mathematics
Massachusetts Institute of Technology
Cambridge, Massachusetts

ACADEMIC PRESS New York San Francisco London 1976
A Subsidiary of Harcourt Brace Jovanovich, Publishers

ACADEMIC PRESS, INC.
111 Fifth Avenue, New York, New York 10003

United Kingdom Edition published by
ACADEMIC PRESS, INC. (LONDON) LTD.
24/28 Oval Road, London NW1

Library of Congress Cataloging in Publication Data

Segal, Irving Ezra.
Mathematical cosmology and extragalactic astronomy.

(Pure and applied mathematics series ;
Bibliography: p.
Includes index.
1. Cosmology. 2. Astronomy–Mathematics.
I. Title. II. Series: Pure and applied mathematics : a series of monographs and textbooks ;
QA3.P8 vol. [QB981] 523.1'2 75-3574
ISBN 0–12–635250–X
AMS (MOS) 1970 Subject Classification: 85A40, 83F05, 22E70, 78A25

PRINTED IN THE UNITED STATES OF AMERICA

Contents

III. Physical theory

IV. Astronomical applications

V. Discussion

Preface

The broad acceptance of the expansion of the universe as a physically real phenomenon has been rooted in part in the apparent lack of an alternative explanation of the redshift. Since its discovery more than a half-century ago, many new observational phenomena have been uncovered, of which quasars and microwave background radiation appear to be particularly fundamental and striking. Nevertheless, there seem to have been few attempts to rework the foundations of cosmology in a way that might tie these phenomena together in a scientifically more economical way. This is probably due more to the momentum of the theoretical studies based on the expansion theory than to its agreement with observation, which has been quite limited and increasingly equivocal.

In this book I present a new theory that is very different from the expansion theory, though equally rooted in the ideas of relativity going back to Einstein, Minkowski, Robb, Veblen, and others. The specific germinal point of the theory was the observation I made 25 years ago that a more natural operator to represent the physical energy than the conventional generator of temporal translation in Minkowski space was a certain generator of the conformal group that physically closely approximates the conventional energy. It has taken a long time to realize that the redness of the observed shift follows from a law of conservation of the new, essentially curved, energy, which necessarily involves a depletion in the old, essentially flat, energy, which is all that can locally be measured and directly observed.

This book is however not merely, or even basically, the presentation of a new model. It is in large part an attempt to lay rational foundations for cosmology on the basis of the most elementary types of causality and related symmetry considerations. It is extraordinary how incisive these qualitative desiderata turn out to be, when integrated with the modern theory of transformation groups. On the purely physical side, the key concepts of time and of its dual energy are given a new precision of definition and treatment that removes much apparent mystery and, in particular, partially mechanizes the murky but important matter of the correlation of mathematical with observational quantities. The title of my original abstract, *Covariant chronogeometry and extreme distances*, summarizes this natural philosophic standpoint, but a corresponding treatment of very small distances (i.e., elementary particles) will require much further exploration.

The new "chronometric" theory emerges in a unique way from this standpoint. It has been interesting to test it against virtually all available relevant astronomical data and to find that, despite its lack of adjustable parameters (other than the unit of distance), it is accepted, in the sense of the theory of statistical hypotheses, by all large or objectively defined samples of galaxies or quasars, indeed at notably high probability levels. In the cases of samples less amenable to rigorous statistical treatment, it typically provides a disdinctly better fit to the data than does the expansion theory, with its free parameter q_0, with one equivocal exception. From an overall scientific point of view, it has been reassuring to find that a fully rational approach to cosmology can lead to physical predictions that conform to observation, and that modern statistical theory is a vital aid in comparing theory with observation, rather than, as appears to be the outlook of some astronomers, an annoying hindrance.

One reason for the delay in promulgating the new theory was that initially one of its predictions appeared in flat contradiction with observation. It implied a square-law redshift–distance dependence for sufficiently small distances, whereas it was "well known" that the relation was observed to be linear. But the mathematical unicity and simplicity of the model, together with its immediate success in dealing with quasars, gave grounds for further exploration of the theory. It has been quite reassuring to find confirmation for the square law in a number of observational studies at moderate redshifts, and overwhelming evidence for a phenomenological square law in the case of low-redshift galaxies. (Of course, actual distances are not directly observable, but the implications of the respective laws for the relations between the redshift, number, apparent luminosity, and angular diameter of luminous objects may be compared with actual observations.) Hubble's original (1929) derivation of the linear law was based on 22 of the more than 700 galaxies included in the low-redshift analysis, and it now appears that the linear law was of a much more tentative character than has been generally realized. The later observations by his associates and successors must be seen in the context of a natural tendency to seek the validation and development of a previously indicated hypothesis, rather than to explore possible alternatives.

I have had useful discussions with many astronomers and mathematicians, and some physicists and statisticians. Particularly valuable assistance was provided by J. F. Nicoll who contributed many corrections to the entire manuscript, assisted in several computations, and kindly permitted the inclusion here of jointly developed material, as noted below. I am also specially indebted to N. S. Poulsen for many corrections to Chapter II and to L. Hörmander, B. Kostant, J. W. Milnor, S. Sternberg, and J. L. Tits for stimulating mathematical comment. I am grateful for the astronomical criticism and information conscientiously given by E. Holmberg, C. C. Lin, D. Lynden-Bell, P. Morrison, G. Setti, B. Strømgren, and others. More formal thanks are due the Universities of Copenhagen and Lund, where the present study was drafted in 1971–1972, the Scuōla Normale of Pisa, the I.H.E.S. of Bures-sur-Yvette, and the University of Warwick, England, where it was continued, and the National Science Foundation where my research was partially supported.

I

General introduction

1. Standpoint and purpose

As mathematical science has evolved, the natural tendency toward the differentiation of labor led to the separation of mathematics and physics, and to the organization of the whole subject along craft and technical lines, rather than along integrated and externally motivated ones. This has resulted in the development of relatively objective and uniform standards of professional work, and its more effective and precise communicability, in mathematics, experimental physics, observational astronomy, and elsewhere. On the other hand, this very useful and natural technical clarification, proliferation, and standardization increases significantly the difficulty of appropriately treating and exposing any single coherent idea or problem of both mathematical and empirical relevance. The present book attempts to steer a sound course between the Scylla of wishful speculation, which may result from attempts to bypass the rigors of conformity to the fundamental scientific disciplines, in the interest of dealing with issues of large relevance, and the Charybdis of elaborate technical refinement, which may result from the determination to ignore questions of relevance, in the interest of achieving an ultimate technical perfection.

This work treats a single coherent conception of space, time, causality, and related notions, and so is presented as an entity, although it depends on the utilization of highly developed areas in the fields of mathematics, physics, and astronomy. Following the present general introduction, it will be treated seriatim from the standpoints of these respective fields; however, a certain underlying terminological and notational uniformity will be employed. The work is primarily a synthesis of foundational developments in each of the fields, coupled with the observation of the existence of a remarkable new space–time model. It involves few special features of unusual technical difficulty; rather, it leads to important and interesting technical problems, by virtue of its crystallization of a new scientific outlook, and its proposal of new notions of time, space, and energy applicable both to cosmology and to microphysics.

Our present notions of these matters are inevitably based on and colored by anthropomorphic perceptions and experiences of them. On the other hand, it is neither inevitable nor desirable that purely theoretical notions of these physically crucial matters should, on the basis of their apparent cogency in anthropomorphic or limited professional realms, be judged binding on theories attempting to deal with the physics of extreme distances. Of course, the subject is inherently very difficult, in that significant observation or experimentation seems possible only on the basis of a substantial theoretical framework, in view of the highly indirect accessibility of objects at these distances; yet any such framework is necessarily rather tentative. A priori, it might well appear doubtful whether a physically conservative and mathematically well-founded treatment of the concepts of time, space, energy, causality, etc. could be sufficiently incisive to attain verifiable quantitative conclusions, and be thereby subject to empirical validation. A methodologically interesting aspect of this book is the demonstration that this is the case.

2. Causality and geometry—historical

When Einstein questioned the absolute nature of simultaneity, and developed a theory of time and communication based on the propagation of light signals, causality considerations were implicitly introduced into the theory of space and time. These emerge more clearly in the work of Minkowski. However, causality was treated in a largely philosophical and intuitive way, as a marginal feature of quantitatively more central matters, e.g., the addition of large velocities. Indeed, the latter feature is identified by Bridgman as the main one in relativity, and one that is logically unrelated to causality. None of these authors, nor their immediate successors, attempted an axiomatic treatment; nor made a consistent explicit separation between mathematical and physical considerations; and to this day (with the exceptions noted below), the notions of "causality," "observer," "clock," and "rod," are commonly used in quite intuitive, if not subjective, senses in work in relativity theory.

The explicit and cogent significance of causality for relativity theory was first recognized and emphasized by Robb (1911–1936) and developed by him into a deductive mathematical theory in which special relativity is effectively derived without any use of such notions as "clock" or "simultaneity" (at different points of space). As recognized by Fokker (1965), Robb thereby founded the subject of *chronogeometry*, in which considerations of temporal order are merged with geometry in a mathematical way, but with a presumption of applicability to physical space–time. A central notion in Robb's theory was a partial ordering in a given space, representing physically the relation of temporal precedence, in the world's space–time medium. Many mathematical axioms, in significant part motivated by optical considerations, with relatively objective physical interpretations (not requiring notions such as observer, clock, or rod at *different* points of space–time) lead after extensive analysis in this theory to the conclusion that the given causality-endowed space is isomorphic to Minkowski space–time, the partial ordering being the usual notion of temporal precedence in this manifold. By modern standards, while Robb's work was quite original and exhibits a high order of mathematical clarity and coherence, it was isolated, unsophisticated, and apparently terminal in intent. Its main significance seems to lie in its formulation of the causality point of view, and demonstration of its power to lead to a more objective and at the same time philosophically satisfying treatment of relativity.

Since the war, the subject of chronogeometry has attracted the attention of a number of mathematicians, including notably A. D. Alexandrov and E. C. Zeeman, and in a modified form, J. L. Tits. In work beginning in the early 1950's, Alexandrov developed a school of work on mathematical relativity, very much from the chronogeometric outlook (explicitly so in Alexandrov, 1967, a key work), which has been contributed to by Busemann (1967) and Pimenov (1970), among others. In 1964 Zeeman rediscovered and exposed cogently the theorem (due originally to Alexandrov and Ovchinnikova, 1953, a work which seems not to have been widely disseminated outside the Soviet Union), that a causality-preserving transformation on Minkowski space is necessarily a Lorentz transformation, within a scale factor. In 1960 Tits, in a key work, published a summary of a classification of all four-dimensional Lorentzian manifolds enjoying certain physically natural transitivity properties.

Chronogeometry has also emerged, in quite a different although related way, from the needs of the general theory of hyperbolic partial differential equations, and our initial acquaintance with the subject was derived from the fundamental work of Leray (1952), which correlated in a very general way the infinitesimal and finite notions of causality. A given hyperbolic equation defines an infinitesimal notion of temporal order, in the form of a proper convex cone in the tangent space at each point of the space–time manifold. Prewar work by Zaremba and Marchaud was completed and applied with cogency in Leray's work. His work, and particularly its chrono-geometric side, has been further developed by Choquet-Bruhat (1971), who has made applications to general relativity; somewhat related work is due to Lichnerowicz (1971). Partially similar but more intricate and specialized ideas have been applied to the problem of the structure of space–time in general relativity by Hawking, Penrose, and a number of collaborators (cf. Ellis and Hawking, 1973), as well as by other recent writers on the problem of singularities in general relativity.

The subject of hyperbolic partial differential equations in the large can be considered in large part as falling under the general heading of causality and evolutionary considerations in functional analysis. This is not a question of pure geometry, of course, but rather of function spaces built on the space–time manifold; nevertheless there are some essential geometrical aspects, and causality plays a crucial role. This is the case, for example, for the key notions of domain of dependence, region of influence, and of causal propagation. Indeed, hyperbolicity may well be necessary as well as sufficient for causal propagation, as evidenced in part by recent work of Berman (1974). This shows in particular that in the Klein–Gordon equation $u_{tt} = \Delta u + cu$ ($c =$ constant), it is impossible to replace $\Delta + c$ by any other semibounded self-adjoint operator in L_2 over space if propagation is to remain both causal and Euclidean-invariant in Minkowski space.

The latter work continues an extensive line of work on the implications of causality for temporally invariant linear operators. The treatment of the dispersion of light by Kramers and Kronig was among the earliest and most influential in this general direction. The work of Bode on the design of wave filters applied a similar idea in a nonrelativistic context, that of linear network theory. Mathematically, the work of Paley and Wiener on complex Fourier analysis, and of Kolmogoroff, and later Wiener, and many others on linear prediction theory, in part relate to causality considerations in a context of temporal development and invariance. The Paley–Wiener theory was extended to a more general setting, applicable to relativistic cases, by Bochner. This was used in the postwar development of the general theory of linear hyperbolic equations due to Gårding and Leray, and thereby connected with causality features.

A partial synthesis of the causality ideas involved in this line of work is involved in the abstract study of linear systems by Fourès and Segal (1955). A general conclusion which is relevant to the present considerations and which emerges from this work is that the "future" may be represented by an essentially arbitrary nontrivial closed convex cone in the underlying linear manifold, without any fundamental loss of scientific cogency in the treatment of global questions. Furthermore, the convexity of the cone is both physically natural and technically crucial.

3. Conformality, groups, and particles—historical

Several different lines of physical mathematics, in addition to the chronogeometric and hyperbolic partial differential equation ones are involved in the present work. Indeed, the model proposed here originated in a study from the vantage points of group deformations and particle classification of the conformal space proposed as a cosmos forty years ago by O. Veblen. Chronogeometry supplied only the decisive final clue and a perspicuous and natural framework.

The rough idea bringing in the deformation of transformation groups was clearly enunciated by Minkowski, who pointed out—admittedly, ex post facto—that the displacement of Galilean relativity by special relativity amounted to a change from one group of transformations to a more sophisticated (and in his view, more attractive) one, of which the first is in a sense a limiting case. This is the limit as the velocity of light becomes effectively infinite, for the phenomena under consideration.† Twenty years later it was found that classical (unquantized) mechanics was similarly a limiting case of a more accurate theory, quantum mechanics. Actually, Planck's constant h, which is involved here, and the velocity of light c, involved in the deformation of the Lorentz into the Galilean group, are fixed constants, unvarying in Nature; but a precise mathematical meaning for the notion of limiting case corresponding to the intuitive physical idea was given in Segal (1951). This concept of group deformation has since been explored in slightly different settings in both the physics and the mathematical literature.

In the light of Minkowski's idea and persistent foundational difficulties in relativistic quantum mechanics, it was natural to raise the question of whether this theory is not in itself a limiting case of a more accurate theory. A model with a discrete space and an associated fundamental microscopic

† Minkowski wrote: "If we now allow c to increase to infinity, and $1/c$ therefore to converge toward zero, we see . . . that the group G_c (the Lorentz group) in the limit when $c = \infty$, i.e. the group G_∞, becomes no other than that complete group which is appropriate to Newtonian mechanics (i.e. the Galilean group). This being so, and since G_c is mathematically more intelligible than G_∞, it looks as though the thought might have struck some mathematician, fancy-free, that after all, as a matter of fact, natural phenomena do not possess an invariance with the group G_∞, but rather with a group G_c, c being finite and determinate, but in ordinary units of measure *extremely great*." ("Space and Time," H. Minkowski, translation of address delivered at 80th assembly of German Natural Scientists and Physicists, Cologne, 21 September 1908; in "The Principle of Relativity," H. A. Lorentz, A. Einstein, H. Minkowski, and H. Weyl, 1923, pp. 78–79; reprinted Dover, New York.)

length, involving a species of approximation of the Lorentz by the de Sitter group, was proposed by Snyder (1947); the set of fundamental dynamical variables did not form a Lie algebra. It was noted by Segal (1951) that the Lie algebras of certain pseudo-orthogonal groups, namely $O(5, 1)$, $O(4, 2)$, and $O(3, 3)$ were deformable into that of the fundamental dynamical variables (momenta, boosts, and space–time coordinates) in relativistic quantum mechanics; a parallel heuristic observation had been independently made by Yang (1947). These Lie algebras were themselves terminal, in the sense that, unlike the Lie algebras of Galilean and special relativity, or of classical mechanics, they were not limiting cases of any other (nonisomorphic) Lie algebras. In physical terms, this indicated a relatively terminal property for a corresponding physical theory, for such a theory based on commutation relations (i.e., a Lie algebra) could be a limiting case of another such theory only if the latter was of higher dimension. While a slight increase in dimension is always a possibility, any large increase would produce many more invariants ("constants of the motion," or mathematically, number of generators of a maximal Abelian subalgebra of Lie algebra, in the relevant representations) than appear compatible with the limited number of states and selection rules observed in elementary particle experiments. Furthermore, groups of larger finite dimension rarely operate effectively on a four-dimensional space–time.

Of the cited pseudo-orthogonal groups $O(5, 1)$ and $O(4, 2)$, the groups of de Sitter and conformal space respectively, have been the most studied. As indicated by Segal (1967a) and Philips and Wigner (1968), $O(5, 1)$ is difficult to reconcile with the fundamental principle of positivity of the energy in quantum mechanics; more specifically, in no nontrivial unitary representation of $O(5, 1)$ does any self-adjoint generator correspond to a nonnegative self-adjoint operator. The group $O(4, 2)$ is free of this failing, and a variety of physical desiderata have pointed to it as a likely candidate for a more accurate higher symmetry group. As essentially the conformal group, it contains the Lorentz group as a subgroup; as shown by Bateman and Cunningham sixty years ago and extended by L. Gross (1964), it is the invariance group of the Maxwell equations—a statement which is mathematically fully meaningful only when Minkowski space is extended to the conformal space treated by Veblen. More recently, experimental indications of scale invariance in elementary particle interactions have led to renewed studies toward the utilization of the conformal group (cf., e.g., Carruthers, 1971).

There have been two major obstacles to the use of the conformal group in foundational theoretical physics, which are roughly macro- and microscopic in nature, respectively. Macroscopically, conformal space is acausal in the sense that at a fixed point x, the limit of the space–time point (x, t) as $t \to +\infty$ is identical with its limit as $t \to -\infty$; these limits exist, the space being closed (i.e., "compact"). This is contrary to physical intuition, leads to serious difficulties of physical interpretation, makes it impossible to distinguish between the advanced and retarded elementary solutions of Maxwell's equations in conformal space, etc. From an elementary particle viewpoint, the fundamental symmetry group is probably more important than the geometrical space serving as particle medium, but the conformal group suffers from a corresponding lack of invariant temporal orientation.

Microscopically, relevance to physical elementary particle observations requires an explicit correlation of representations, and a set of generators of a maximal abelian subalgebra of the enveloping algebra, with observed particles and their quantum numbers. This is a highly vertical and complex process; relatively small differences in the initial aspects of this correlation may ramify and produce gross differences in the implications subject to empirical validation. For example, it is not clear a priori whether the energy and other conventional dynamical variables should remain unchanged, as is possible because of the inclusion of the Lorentz group in the conformal group, and is assumed implicitly in most of the theoretical physical literature (but which leads to difficulties because of the lack of conformally invariant wave equations for massive particles, among other reasons); or whether the energy, etc. require modification so as to involve the full conformal group in a more essential way, as originally proposed by Segal (1951). In Segal (1967a), qualitative evidence for such a new generator was adduced: (a) unlike the conventional generator P_t, which cannot lead to mass splitting, according to a theorem due in infinitesimal form to O'Raifeartaigh (1965) and in global form to Segal (1967b), the new energy operator P_t' (which corresponds to a generator of the conformal group which is essentially different from, i.e., nonconjugate to, $\partial/\partial t$) may have a discrete spectrum; (b) the idea that temporal displacement, as dynamically fundamental, should be definable in a mathematically unique and natural way is substantiated by P_t', which has such definitions, in terms of $O(4, 2)$ as the generator of the corresponding $O(2)$ subgroup, and in terms of the twofold covering group $SU(2, 2)$ as the correspondent to its simplest generator. However, a quantitative check on the validity of this definition via microscopic observation appears difficult except in conjunction with a number of additional assignments or correspondences between apparent quantum numbers and theoretical operators, required to identify the particles whose energy spectrum should be correlated with an appropriate representation of P_t'.

In Segal (1971) it was observed that the acausality of conformal space–time could be remedied in the present connection through its replacement by its locally identical universal covering space; this covering has an infinite number of sheets, and is thereby suggestive of large-scale macroscopic phenomena, e.g., those of large-distance astronomy. Theoretical exploration of this infinite-sheeted covering space from the standpoint of more objective notions of observer, clock, and rod, in the conservative spirit of the foundations of geometry, leaves little doubt that the appropriate notion of time is different in the large from the special relativistic one, although microscopically nearly identical to it. This new time τ is identical with that with the new energy P_t' just mentioned is associated (i.e., $P_t' = -i(\partial/\partial\tau)$ essentially); it leads directly to physical implications which can be checked against observation—in astronomy, rather than in microphysics. This book details the basic theory involved; the astronomical implications; and their successful and interesting confrontation with observation.

4. Natural philosophy of chronogeometric cosmology

As earlier indicated, when the underlying space–time is linear (i.e., a linear vector space), it is rather well established, in a variety of ways, that an appropriate general starting point for a notion of causality is a given closed convex cone in the space–time manifold, representing physically the "future."

The general process of nonlinearization of a theory, that is, the transference to an arbitrary sufficiently regular n-dimensional manifold of a theory established for n-dimensional vector spaces, then suggests as a starting point for causality considerations in a (nonlinear) manifold a structure consisting in the assignment to each point of a closed convex cone in the tangent space at the point.

In physical terms, this is the specification of infinitesimal future, i.e., the set of all future directions at the point.† A given linear hyperbolic partial differential equation provides a particular such assignment, which we shall call a causal orientation. However, from foundational and philosophical viewpoints, there is no particular reason to assume that the causal structure of space–time arises in this way from a hyperbolic equation; rather, hyperbolicity should be an expression of compatibility of propagation with the given causal orientation.

It thus appears—from other standpoints as well—that a natural starting point for the study of temporal order and associated developments consists of a smooth manifold together with a causal orientation, in the sense

† This specification can be regarded as a mathematical formulation of "time in its most primitive form," in the sense of Maxwell, who wrote: "The idea of Time in its most primitive form is probably the recognition of an order of sequence in our states of consciousness." ("Matter and Motion," London, 1877, reprinted Dover, New York.) One of our aims will be to show that in space–time manifolds with realistic features, this apparently minimal physical structure already suffices to determine much of the physical interpretation—the notions of clock, rod, energy, momentum, etc.

of a smooth assignment to each point of the manifold of a nontrivial closed convex cone in the tangent space at the point. Initially, such a causal orientation might appear too qualitative for technical cogency, in comparison with the familiar differential-geometric structures. However, the notion of causally oriented manifold appears to be one of considerable economy and naturalness for the analysis of temporal order, both from a philosophical and a mathematical standpoint.†

All this is not to say that it would not be interesting or possible to have a treatment of causally structured spaces which did not depend on the local smoothness of the space. (For example, there is no essential difficulty in extending the notion of causal orientation to the genre of arbitrary topological spaces.) But until one has a better understanding of causality matters in the more accessible context of smooth manifolds, it might well be mathematically foolhardy as well as physically irrelevant to attempt to obtain results for such general spaces comparable to what may be expected to be available in the smooth manifold case.

Indeed, even the concept of causally oriented manifold is highly qualitative, from a physical standpoint. Although timelike and spacelike directions in the manifold are determined, the notions of time and space, in the precise senses associated with the ideas of "clock" and "rod" are elusive in this context. It is difficult to see how physically cogent results can be obtained without a "clock," or an equivalent structure. For in physics one observes, largely, the change in the state of a system from one time to another. To give meaning to a statement concerning the change of state, one needs an objective parametrization of states which is time-independent, in addition to an objective notion of time itself. Moreover, the key physical notions of energy and scattering appear uniquely and effectively definable only when there is a notion of temporal invariance.

In an arbitrary causally oriented manifold, there may well be many different types of "world lines" (mathematically, maximal chains relative to temporal precedence as order relation); and different, possibly topologically distinct or causally inequivalent spacelike surfaces (i.e., submanifolds such that neither of any two of its points precedes the other, and which are maximal with respect to this property). The notion of a clock as an additive functional on intervals of world lines is conceptually acceptable, but is much

† One rough indication of the cogency of a causal orientation is the existence of evidence for, and the lack of evidence against, the conjecture that the automorphism group of a causally oriented manifold is finite dimensional (i.e., a Lie group in the classical sense), provided the cones in question are proper. Another is evidence that an analogue to the Alexandrov–Ovchinnikova–Zeeman theorem holds; any one-to-one transformation of a *globally causal* manifold of the indicated type is automatically smooth, and so a causal automorphism (cf. Choquet-Bruhat, 1971).

too limited to provide an adequate basis for correlation of the theory with empirical physics. The stronger assumption, that a hyperbolic pseudo-Riemannian structure is given in the space–time manifold, is likewise insufficient, e.g., to determine a fully viable notion of energy, whose precise definition and nature is still controversial in general relativity.

A conceptually natural way to introduce notions such as observer, clock, and rod, a way which generalizes special relativity and is closely related to elementary particle considerations, is to assume and exploit group invariance properties. The plausible and widespread if partially implicit idea that a certain temporal stability underlies the possibility of describing the dynamics of real physical systems is appropriately formalized by the assumption that the causal manifold in question admits a nontrivial class of "temporal displacements," these being automorphisms of the manifold (qua causal manifold) which carry each point into one which is either later or earlier than the given point. *Clock* may correspondingly be defined, essentially as a continuous one-parameter group of such temporal displacements. *Time* is then uniquely determined, within a scale factor, as the additive parameter t of this group, normalized (partially) so that $t > 0$ corresponds to a forward displacement (i.e., one carrying each point into a later point). Given any such clock, one may of course construct other clocks by conjugating the given one by other automorphisms of the causal manifold; physically, any such automorphism leaving a point fixed can be interpreted as a change in the frame of reference of an observer at the point. In a similar way the important although less fundamental notion of *rod* can be associated with the assumed homogeneity and isotropy of *space*. *Observer* then corresponds, in operational terms, to a splitting of the space–time manifold into respective space and time components, in such a way that the groups of the "clock" and "rod" act only on the corresponding component, the temporal action T_t being simply the transformation $\tau \to \tau + t$. Quantum mechanics relative to a given causally oriented manifold is naturally taken to involve a representation of the fundamental symmetry group of all causal automorphisms, i.e., causality-preserving transformations on the manifold. The *energy* for a given observer is then definable as the infinitesimal generator of the corresponding one-parameter group representing his clock. The *spatial momenta*, generalizing the usual linear and angular momenta, are correspondingly describable in terms of the generators of the "spatial displacement" group, consisting of those causal automorphisms that affect only the space component of the observer. Spatial homogeneity means that this group acts transitively; spatial isotropy means that the group acts transitively on the directions at any point. The assumptions of temporal and spatial homogeneity, and of spatial isotropy, are tantamount to the conservation of energy, of linear and angular momentum, respectively. Without these laws, the correlation of theoretical and empirical physics as they exist today would appear impossible.

The physical validity of these notions is confirmed by the observation that in special relativity the usual notions of observer, clock, etc. are in essential conformity with the general ones indicated here. Moreover, it follows from the structure of the causal automorphism group of Minkowski space earlier indicated that there are no other observers or clocks; thus, all observers are conjugate within the Lorentz group augmented by the group of scale transformations. Physically, it seems clear that the Cosmos is four dimensional, and that absolute simultaneity does not exist, i.e., no mode of communication or interaction has infinite velocity. Mathematically, these are readily formulated, the first assumption without change, the second as the assumption that the future cones contain no full straight lines (as they do in primitive Newtonian mechanics). Together with the existence of an observer, these seem to form a physically conservative and intuitive set of axioms for the Cosmos.

It should be interesting to determine all mathematical cosmos in this sense, particularly those which approximate locally, in the vicinity of a point, the Minkowski cosmos. But already a certain ambiguity in the Minkowski cosmos itself appears. While globally all observers are equivalent, locally this is not the case; the concepts of local observer, time, space, etc. can be introduced in entirely parallel fashion to the global concepts, by replacing the global transformation groups involved by local ones. The theoretical concept of local observer seems physically quite relevant since direct measurement of the entire Cosmos is impossible. Indeed, in Minkowski space, considered as a causal space–time continuum, there exist invariant local observers that are nonconjugate to special relativistic ones; and these local observers are applicable to regions which when scaled in accordance with physical parameters are far larger than those accessible to direct observation. The question of which of the local observers is physically correct is a real one; it cannot be eliminated by a mathematical transformation; while subject to various theoretical considerations, it must ultimately be weighed against observation, as we shall do later.

The question arises in particular of whether the same clocks are appropriate, in the sense of yielding a convenient notion of energy, including energy conservation, etc., at all distance levels of physics (or for all types of interactions). The conventional standard relativistic model is very well established at the middle distance levels, but its applicability at the extremes (i.e., extragalactic and fundamental particle physics) is largely a matter of extrapolation in the absence of any other established theory. Since dynamical theories primarily describe transitions from one approximate stationary state to another, such states at the middle distance level may appear complex in terms of inequivalent observers, and in particular nonstationary; conversely, simple descriptions of systems at the extreme distance levels may depend on the analysis of their dynamics in terms of states that are approximately stationary relative to clocks nonconjugate to any Minkowski clock. This abstract possibility is exemplified in the treatment of cosmology later in this book, in terms of the model briefly indicated in the next section.

All of the foregoing has been independent of dynamical assumptions, apart from the implicit one that observed fields and particles are appropriately described by functions defined on the Cosmos, with values in a suitable spin space; and that the equations determining temporal development should imply compatibility with the causal orientation in the Cosmos (in particular, finite propagation velocity), as well as enjoy invariance with respect to the causal automorphism group (or at least the subgroups earlier indicated). These requirements probably essentially imply that the dynamical equations should be partial differential equations which are hyperbolic relative to the given causal orientation (cf. the related discussion earlier).

5. The universal cosmos—sketch

There exists a cosmos that is locally identical to the Minkowski cosmos, and has a certain theoretical universality, in being apparently applicable in a fundamental sense at all distance levels. It may be described as the universal covering space of the conformal compactification of Minkowski space. For these reasons, and by virtue of applications made below to large-distance astronomy, it seems appropriate to designate this model as the *universal cosmos*. Its essential ideas were summarized in a preliminary account in Segal (1972).

The mathematical origin of this cosmos may be briefly indicated as follows. As earlier indicated, conformal space $\bar{M}$, obtained from Minkowski space M roughly by the adjunction of a light cone at infinity, is highly symmetrical, but is acausal. Being non-simply-connected, it admits nontrivial coverings, which are locally identical to $\bar{M}$, and hence locally Minkowskian. The finite coverings of $\bar{M}$ are likewise acausal, but the universal covering manifold $\tilde{M}$, is globally causal with respect to its inherited causal orientation, and defines an admissible mathematical cosmos. The space–time conformal group G acts only locally on $\tilde{M}$, but its universal covering group $\tilde{G}$ operates globally on $\tilde{M}$. Both the covering of $\bar{M}$ by $\tilde{M}$ and that of G by $\tilde{G}$ are infinite-sheeted, and indeed the group $\tilde{G}$ is not a linear group. The center of $\tilde{G}$ is $Z_2 \times Z_\infty$; and the Z_∞ component precludes a faithful finite-dimensional linear representation; however $\tilde{G}/Z_2$ acts faithfully, as a group of conformal transformations, on $\tilde{M}$.

The universal cosmos $\tilde{M}$ is locally identical chronogeometrically to

Minkowski space, and is essentially† the only other cosmos with this property enjoying physically natural symmetry and causality properties. Its validation as a realistic model is discussed below in terms of quantitative applications at the extragalactic level of distance. The physical interpretation is fixed by the distance scale, which may be determined from redshift measurements; a convenient equivalent physical constant may be described informally as the radius R of the universe. Three fundamental physical units are determined in a geometrical way; this is impossible in Minkowski space. $\tilde{M}$ is invariant not only under the 11-parameter Lorentz group extended by scale transformations which acts on Minkowski space, but the full 15-parameter conformal group (more precisely, universal covering group thereof); and any two global physical observers are conjugate with respect to this group. The one-parameter subgroup of this group representing temporal evolution—again unique within conjugacy—is essentially distinct from, i.e., nonconjugate to (within the causal automorphism group) that in special relativity. However, as $R \to \infty$, the universal cosmos deforms locally into Minkowski space, and the universal covering group of the conformal group deforms correspondingly into the Lorentz group together with scale transformations, four of the generators deforming into zero; and arbitrarily large bounded regions in Minkowski space can be approximated arbitrarily closely by the universal cosmos, by taking R sufficiently large. In particular, the universal energy deforms into the special relativistic energy as $R \to \infty$. One thus obtains a particularly concrete form of deformation of one Lie group into another, involving in addition a type of deformation of certain representations of one group into representations of the other.

For any given global observer O on the universal cosmos, and any point P in the cosmos, there is a unique local relativistic observer O'_P (and corresponding Lorentz frame) which is tangential to O at P; and O and O'_P agree near P within terms of third and higher order in the distance from P. Thus O'_P is locally nearly P-independent; however, if Q is remote from P, O' and O'_P are physically quite different; their Lorentz frames are related by the product of a scale transformation with a Lorentz transformation. In particular, the Lorentz frame of Q is in motion relative to that of P, which may be regarded as a virtual Doppler effect; however, from the standpoint of the globally more fundamental universal time, the situation is static.

† An open orbit in $\tilde{M}$ under the action of $SO(2, 3)$ enjoys the most crucial properties; but the regions of influence of compact regions in space may ultimately become noncompact. In any event, the predicted relations between the primary observable quantities (redshifts, magnitudes, number counts, etc.) would not differ from those for $\tilde{M}$. The orbit decomposition of $\tilde{M}$ under $SO(2, 3)$ was determined by B. Kostant, who noted also the existence of invariant Lorentz metrics in each of the two open orbits. The global causality of this space was noted by Wigner (1950), and it is naturally included in the list of Tits (1960).

The natural energy operator $-i(\partial/\partial\tau)$ for the universal cosmos is however not scale covariant†; built into its structure is a fundamental length, the radius of the universe. On the other hand, measurements of microscopic phenomena taking place wholly within a laboratory (i.e., excluding gravitation and redshifting phenomena) are so far as is now known, and may well be in fundamental principle, *scale covariant*, i.e., based on units of time and distance that are wholly conventional. This would suggest that the observable representing local measurement of frequency is not $-i(\partial/\partial\tau)$, but rather the conventional, scale-covariant operator $-i(\partial/\partial t)$, which as it develops is precisely the scale-covariant component of $-i(\partial/\partial\tau)$, and locally unobservably different from it (in natural units). Despite the very small local difference between these operators, their noncommutativity in the large implies that the relativistic energy operator is not at all conserved under universal propagation over a lengthy period of time. It is in fact a purely mathematical deduction that the apparent frequency of light, propagated in accordance with Maxwell's equations by universal-time displacements, is shifted to the red.

6. The chronometric redshift theory

More specifically, the redshift z is found to vary with the distance r from the point of emission in accordance with the law: $z = \tan^2(r/2R)$, where R is the "radius of the universe." This is in itself not a relation between observable quantities; but a variety of relations between observed quantities, such as redshifts, apparent luminosities, number counts, and apparent angular diameters, are readily deduced from this law. These predictions from the theory have been found to be in much better agreement with actual raw observational data than would a priori have been expected for an astronomical theory. Some of the outstanding predictions, and their relations to observation, are as follows:

(1) For small r, z varies as r^2, in accordance with the observational analyses of Hawkins (1962) and G. de Vaucouleurs (1972), and as preferentially indicated by the complete sample of radio galaxies due to Schmidt (1972c), the list by Arakelyan *et al.* (1972) of Markarian galaxies at substantial redshifts, a sample of Seyfert-like Markarian galaxies studied by Sargent (1972), a small sample of N-galaxies treated by Sandage (1967), and rather definitely, the large sample of G. and A. de Vaucouleurs (1964). The very good fit of the chronometric theory to the m–z data for the de Vaucouleurs

† Analytically, a generator X of the fundamental symmetry group is scale-covariant if $[X, K] = X$, where K is the infinitesimal generator of scale transformations, $K = \sum_j x_j(\partial/\partial x_j)$.

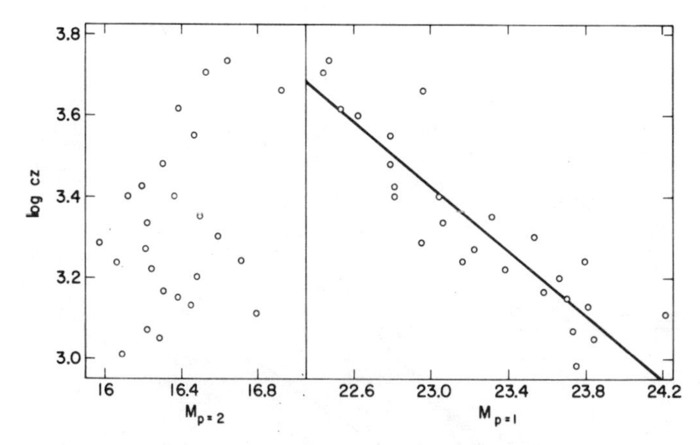

Figure 1 *The redshift–absolute magnitude relations for the tenth brightest galaxies in bins of size* 20 *galaxies, ordered by redshift, included in the de Vaucouleurs tape.*

All galaxies having m–z–θ data, 742 in all, were used. The absolute magnitudes for a square-law redshift–distance dependence, shown on the left, differ negligibly from those based on: (a) the maximum-likelihood power law fitted to the data; (b) the chronometric theory. Those shown on the right for the linear redshift–distance law have a trend that differs imperceptibly from that predicted by the chronometric theory, of slope 2.5, which is shown here as a solid line.

galaxies, together with the strong trend of the deviations from the expanding-universe hypothesis, is shown by Figure 1.

The apparent deviation from the law of the sample of bright cluster galaxies studied by Sandage is probably primarily a selection effect. This is evidenced by the extremely irregular $N(< z)$ distribution of this sample; this distribution is moreover highly deviant, even for $z < 0.04$, both from that to be expected in an expanding universe and the observational relation for all such galaxies with published redshifts, as compiled by Noonan (1973). Some of these circumstances are shown in Figure 2. It is evidenced also by an apparently very large dispersion in the intrinsic sizes of the galaxies. No objective statistical criterion for the sample has been published, and in fact its origin appears to be lost in early decisions of Humason. In addition, the deviation is augmented by the model-dependency of the apertures of observation which introduces a z-dependent trend, and possibly by the inherent tendency of an established theory to influence difficult observations.

(2) The apparent magnitude V depends on redshift z according to the relation

$$V = 2.5 \log z - 2.5(2 - \alpha) \log(1 + z) + c,$$

where α is the spectral index and c is a parameter representing the intrinsic luminosity of the source; corrections for absorption, aperture, intrinsic

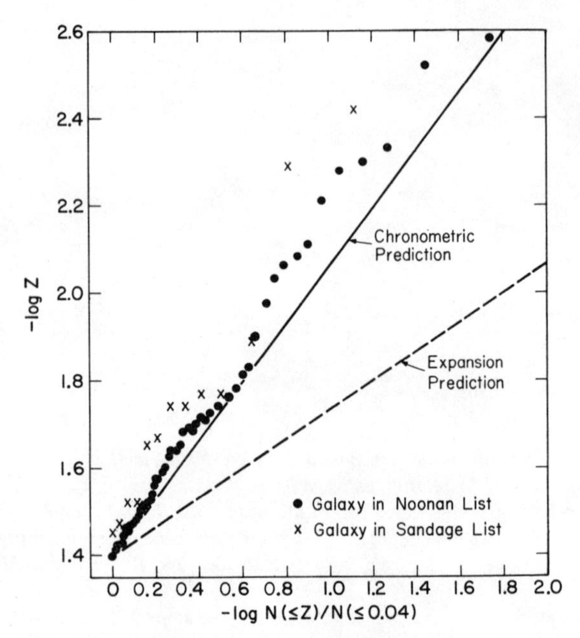

Figure 2 *The N–z relation for two samples of brightest cluster galaxies, in the range* $z < 0.04$.

Shown are 56 galaxies from Noonan (1973) and 13 galaxies from Sandage (1972b), Table 2. The $N(< z)$ curve for the Sandage sample differs even more from the expansion prediction than from the Noonan curve, even for very small redshifts. The respective values of $\partial \log N(< z)/\partial \log z|_{z=0}$, which should be unaffected by observational magnitude cutoff by virtue of the evaluation at $z = 0$, are ~ 1 for the Sandage sample, 1.45 for the Noonan sample, 1.5 for the chronometric prediction, and 3.0 for the expansion prediction.

motion of the source, if any, are not included. In particular, as z increases from 0.4 to 4.0, V should increase by ~ 1.1 mag (for $\alpha = 1$; for $\alpha = 0.7$ the increase is ~ 0.7 mag), as contrasted with the increase of ~ 5 mag on a typical expansion-theoretic hypothesis.

Quasar observations are in excellent agreement with the new law, with an overall dispersion for all quasars of less than 1 mag, and of 0.3 mag for the "locally brightest" fifth of the quasars, where "local brightness" is a model-independent measure defined as the difference of the magnitude of the object and the average magnitude of the six quasars at the nearest redshifts (three at greater and three at lower redshifts than that of the object). These dispersions are much less than those from the Hubble law (by more than a factor of 3 in the case of the last sample, of 32 quasars). In all substantial previously identified samples of quasars, including complete samples of radio sources due to Schmidt (1968) and Lynds and Wills (1972) and of optically selected quasars due to Braccesi *et al.* (1970), the chronometric dispersion is less than the Hubble-law dispersion, generally by factors of the

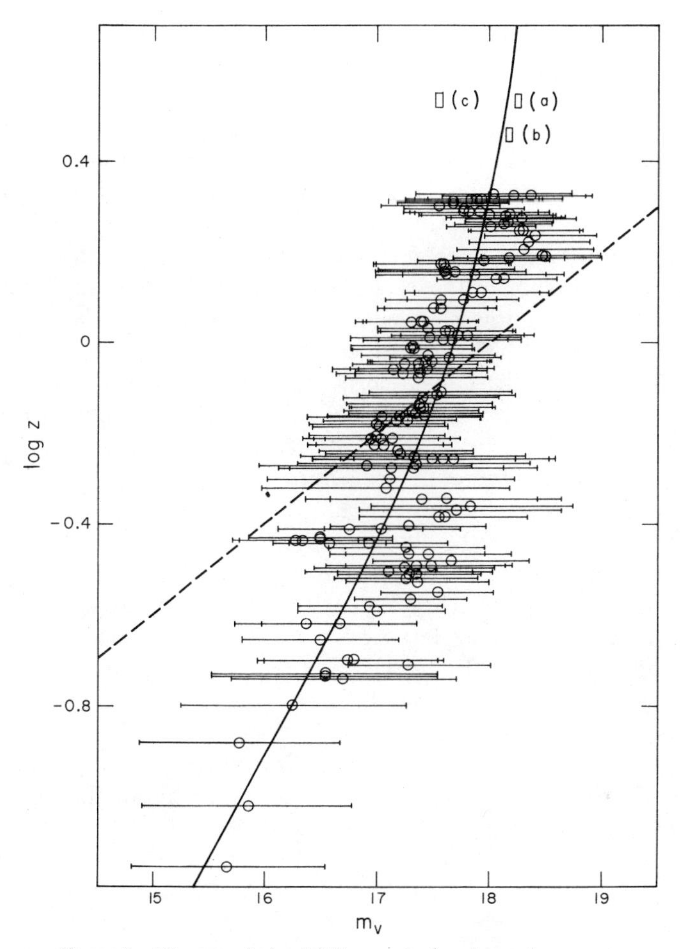

Figure 3 *The smoothed redshift–magnitude relation for quasars.*

○, average magnitude of seven quasars at approximately the same redshift. The bar indicates the unbiased probable error of the group. Chronometric (—) and expansion (---) ($q_0 = 1$) predictions, with zero points fitted to the average magnitude of the sample, include all quasars in the list of DeVeny *et al.* (1971) having unquestioned data. The three quasars of maximal presently published redshifts are denoted as (a), (b), and (c). On the expansion theory, (c) (at redshift 3.53) is $\sim$ 50,000 times as bright as the average nearby galaxy, and is moving with $>$ 0.9 times the velocity of light. On the chronometric theory, (c) has about the same intrinsic brightness as nearby bright galaxies, and need not be moving at all.

order of 2. This is true also of radio and infrared luminosities, when available. For the Einstein–de Sitter model, the disparity in dispersion is typically $\sim 10\%$ greater. The nearly optimal fit of the chronometric theory to the uncorrected quasar data, together with the pronounced deviation of the Hubble law, is shown by Figure 3.

(3) The redshift distribution of quasars that would be expected from a spatially uniform distribution in the present model conforms extremely well to observed redshift distributions, whether for the complete list of known quasars given by DeVeny *et al.* (1971), for complete samples limited in various magnitudes, or over specified redshift intervals. In contrast, in the expansion theory, strong evolutionary effects are required to explain the distribution, as documented by Schmidt (1968, 1972b), Lynds and Wills (1972), and others, and further strongly confirmed in the present analysis employing systematically the Kolmogorov–Smirnov test for a variety of substantial quasar samples, as well as Schmidt's V/V_m test when applicable. The latter test rejects the hypothesis of spatial homogeneity for the Peterson and de Vaucouleurs lists of galaxies on the expansion hypothesis, but accepts the chronometric hypothesis at substantial probability levels.†

(4) Perhaps the major anomaly in quasar phenomena from the expansion-theoretic standpoint, their apparent unprecedently large energy output, is fully resolved by the change in the relative luminosities of quasars and galaxies implied by the chronometric theory. When analyzed on this basis, observations indicate that the average quasar is $\sim$0–1 mag fainter than the average brightest cluster galaxy, and quite comparable to an average N-galaxy, or average Seyfert galaxy. The hypothesis that quasars are the cores of certain bright galaxies whose outer regions are not seen at larger redshifts is somewhat supported by this and other consequences of the chronometric hypothesis.

(5) A second major anomaly regarding quasars, the apparent relative cutoff in quasars above $z \sim 2.5$, is well resolved without the use of fairly drastic hypotheses, hardly subject to independent verification, required to this end in the expansion theory. Specifically, the theory predicts that for any object uniformly distributed in space, the expected number in the redshift range $2.25 < z < 3$ will be $\sim 8\%$ of that in the redshift range $0 < z < 2.25$. In the light of probable and partially documented changes in the spectra of quasars at higher frequencies and spectroscopic selection effects making their identification more difficult at higher redshifts, this is quite compatible with quasar observations; the corresponding figure of $\gtrsim 40\%$ on the unembellished expansion theory is not.

(6) Apparent superluminal velocities of large redshift objects are eliminated on the chronometric hypothesis by the reduction in the theoretical distance to large redshift objects, typically by an order of magnitude.

† The Peterson galaxies were observed at given expansion-theoretic apertures, and their magnitudes consequently require correction to chronometric apertures for a valid test of the chronometric hypothesis, just as they must be used as given for a test of the expansion hypothesis. The de Vaucouleurs galaxies were measured at apertures determined by observational rather than theoretical criteria, and were used in the statistical analysis without correction for both hypotheses.

(7) The $N(m)$ relation for quasars is very well fitted by the chronometric curve for a single luminosity class, convolved with a normal law luminosity function of dispersion equal to that observed. This explains the apparent cutoff in quasar identifications at faint magnitudes ~ 20.5 noted by Bolton (1969), Braccesi *et al.* (1970b), and others, a phenomenon that appears anomalous from the expansion-theoretic standpoint. The expansion-theoretic $N(m)$ relation for a complete sample is in fact in disagreement with the observational relation for quasars in the DeVeny list even when limited to relatively bright magnitudes such as 18.0.

(8) The index $-\partial \log N/\partial \log S$ for a single luminosity class and spectral index $\alpha < 1$ is ~ 1.5 for very bright sources but eventually becomes infinite as S decreases, according to the chronometric theory, following which it drops to zero. When convolved with a luminosity function of about one decade width, the theoretical curve rises quite moderately above 1.5 for fairly bright sources and eventually declines to a value $\lesssim 1$, in qualitative agreement with observations of Pooley and Ryle (1968) and Kellermann *et al.* (1971).

(9) Predictions regarding the angular diameter in relation to redshift are in satisfactory agreement with the data on double radio source quasars as given by Miley (1971). The angular diameter measured here is properly identifiable with the metric diameter treated theoretically, and all dispersion in the observations attributable to variation in the redshift, as measured in a model-independent fashion, is removed by the chronometric relation. The situation is similar as regards double radio galaxies listed by Legg (1970).

The theoretical deductions involved here are obtained in a quite direct and objective manner, and involve no free parameters, other than the distance scale, which is determined by R. Further observational confirmation and predictions, standard statistical significance tests, and a discussion of corrections and selection effects, are given in detail in Chapter IV.

7. Theoretical ramifications; the cosmic background radiation

We close this section with comments on theoretical aspects which seem likely to be raised in the minds of certain groups of readers. First, on the general mathematical side, the question arises of whether the methods involved here are ad hoc and entirely particular, or whether the theory can be understood as an individual instance of a general type of theory. Indeed, the latter is the case. There is an analogous theory for general classes of causally oriented manifolds, in relation to corresponding flat manifold tangential to them. The Cayley transform being causal, there is no chronogeometric local distinction between the two different manifolds, but in the large there are topological and other differences. Our basic physical assumption is that local measurement (e.g., of frequency) is in terms of the flat tangential causally oriented manifold—roughly that microscopic observation is based on a Minkowski clock; but that true free temporal evolution is as given on the global curved manifold, i.e., runs on the universal clock. Thus, the universal energy of a free wave or particle is conserved, but the apparent frequency of a photon emitted from an atom changes noticeably after a long time because it is stationary with respect to the Minkowski rather than universal clock. Such a limitation on local phenomena and measurement is a priori plausible because of the absence of any absolute distance scale for measurements of entirely microscopic phenomena. Normalization of the commutation relations of quantum mechanics fixes the values of $\hbar$ and c as unity but leaves unspecified one of the fundamental units. In the universal cosmos the natural convention $R = 1$ fixes the distance scale and completes the specification of units.

The causal manifolds involved are all globally hyperbolic and have defined on them analogues of Maxwell's and Dirac's equations. In addition they are extremely symmetrical, being universal covering manifolds of Shilov boundaries of classical Siegel domains, whose automorphism groups are closely connected with the presently relevant physical symmetry groups. In the case of dimension 4, however, there is essentially only one known instance of the general theory, viz. the universal covering manifold of the conformal compactification of Minkowski space in relation to Minkowski space (or equivalently, the universal covering group of $U(2)$ in relation via the Cayley transform to the Lie algebra of $U(2)$ as identified with the 2×2 Hermitian matrices with their usual ordering). This seemingly purely mathematical aspect has in our view a certain physical relevance, in diminishing the selection effect involved in formulating any theory designed to explain previously observed phenomena, and thereby enhancing the significance of whatever agreement is found between theory and experiment. Indeed this is in essence no more than the broadly recognized distinction between correlating data by curve-fitting and the like, and the formulation of a true theory based on general ideas and principles.

Second, the relation to dynamical theories—general relativity, the question of the origin and "age" of the universe, elementary particle dynamics—is likely to be raised. Since conformal space is conformally locally identical to Minkowski space, the present model for space–time stands in essentially the same relation to general relativity as does special relativity. As a variant of special relativity, it is essentially a purely kinematical structure, on which one is free to impose interactions as in the case of Minkowski space. In particular, the ideas of general relativity carry over bodily and its applications to local gravitational phenomena (e.g., within a galactic cluster) appear unaffected.

On the other hand, material dynamical content resides in the postulate concerning local observations of dynamical quantities, to the effect that these are represented not necessarily by generators of true, global symmetries, but rather by generators of corresponding symmetries in the tangential flat model. While angular momenta, for example, are unchanged, the energy and linear momenta are altered in essential ways. The true energy is no longer represented by $-i(\partial/\partial t)$, but by an operator which while extremely simple and natural from the standpoint of universal space, appears complicated in terms of Minkowski space. It may be put in the form $-i\,\partial/\partial\tau$, where τ is the universal time. This differs from $-i\,\partial/\partial t$ by an operator that is virtually negligible up to galactic distances, and so as an interaction Hamiltonian should not be responsible for any readily observable microscopic processes. Moreover, as the radius of the universe becomes infinite, this interaction operator $i(\partial/\partial t - \partial/\partial\tau)$ tends to zero, in accordance with Minkowski's concept of limiting case. It is relevant to note that the difference between the universal and special relativistic energies, the "superrelativistic energy," is represented by a positive Hermitian operator in all physical (hence positive-energy) representations of the fundamental symmetry group (for example, in the representation defined by Maxwell's equations).

The redshifting process may be regarded in the chronometric theory as a conversion of relativistic into superrelativistic energy, which inevitably accompanies the delocalization of a photon wave function, the superrelativistic energy being negligible for a localized photon. The conversion becomes in classical theory total at redshift $z = \infty$, but at low frequencies and high redshifts the quantum-theoretic dispersion in frequency (which arises from the noncommutativity of the operators representing the relativistic and superrelativistic energies) will significantly broaden the spectrum of the radiation. It should then appear as background radiation, the totality of which would be in a state of equilibrium, assuming that the temporal homogeneity of the universal cosmos is dynamically as well as kinematically valid. By conservation of energy and maximization of entropy, this radiation should have a blackbody spectrum, as is consistent with observations of the microwave background, which is thereby theoretically predicted.

II
Mathematical development

1. Causal orientations

Unless otherwise specified, all manifolds will be taken as real, finite dimensional, and of class C^∞; and all groups as finite-dimensional Lie groups; there is, however, no essential difficulty in being considerably more general.

Definition 1 A *convex cone* in a real linear space **L** is a subset **C** with the property that if $x, y \in \mathbf{C}$, and if $a, b \geqq 0$, then $ax + by \in \mathbf{C}$. Such a cone is *trivial* if either $\mathbf{C} = \mathbf{L}$ or $\mathbf{C} = \{0\}$; otherwise it is *nontrivial*. It is *proper* if it contains no full straight lines and is not a direct product of a ray and a cone of codimension one. It is C^∞ (respectively analytic) if there exists a finite set Φ of C^∞ (respectively analytic) functions on **L** such that $x \in \mathbf{C}$ if and only if $f(x) \geqq 0$ for all $f \in \Phi$.

Definition 2 An *infinitesimal causal orientation* in a manifold M is an assignment $p \to C(p)$ to each point p of M of a nontrivial closed convex cone $C(p)$ in the tangent space at p, which is locally definable by a finite number of inequalities on continuous functions of p and the components of tangent vectors. Specifically, this means that each point p has a neighborhood N in which there exist local coordinates $x_1, \ldots, x_n$ and a finite set Φ of continuous functions on $N \times R^n$ such that if $q \in N$ and $\lambda \in T_q$ (the tangent space at q), then $l \in C(q)$ if and only if $l = \sum_k a_k(\partial/\partial x_k)|_q$, and $f(q, a_1, \ldots, a_n) \geqq 0$ for all

$f \in \Phi$. If M is C^∞ (respectively analytic or algebraic), the causal orientation is called C^∞ (respectively analytic or algebraic) if it is similarly defined by a finite number of C^∞ (respectively analytic or algebraic) functions.

Example 1 Let G be a Lie group, let $\mathbf{G}$ denote the Lie algebra of all right-invariant vector fields on G, and let $\mathbf{C}$ be a nontrivial closed convex cone in $\mathbf{G}$. Defining $C(p) = [X_p : X \in \mathbf{C}]$ is easily seen to give a causal orientation in G, which is C^∞ (respectively analytic) if the cone $\mathbf{C}$ is such.

Definition 3 An infinitesimal causal orientation in a group G is said to be *right-invariant* if of the form given in Example 1; *left-invariant* if the same except that left-invariant vector fields are employed; and simply *invariant* if both right and left invariant.

Example 2 Let $G = U(n)$, i.e., the group of all unitary $n \times n$ complex matrices. Then $\mathbf{G}$ can be identified with the linear space $\mathbf{H}(n)$ of all $n \times n$ complex Hermitian matrices, via the isomorphism: if $X \in \mathbf{G}$, then $X \to H$, where $H \in \mathbf{H}(n)$, if for all C^∞ f on $U(n)$, $Xf = (d/dt)f(e^{itH}x)|_{t=0}$ (where x is a bound variable); and to avoid undue circumlocution, we shall use this identification. Let $\mathbf{C}$ denote the subset of $\mathbf{G}$ consisting of those H in $\mathbf{H}(n)$ for which $H \geqq 0$, in the usual sense that $\langle Hx, x\rangle \geqq 0$ for all vectors x, $\langle \cdot, \cdot \rangle$ denoting the usual complex Hermitian positive definite inner product in C^n. Then $\mathbf{C}$ is a closed convex algebraic cone in $\mathbf{G}$. It is evidently invariant under the action (induced) on $\mathbf{G}$ of inner automorphisms of G, and so defines an invariant infinitesimal causal orientation. It is not difficult to verify that if $n = 2$, this is the only quadratic such structure, apart from those defined by $-\mathbf{C}$, and by the one-dimensional cones $[lI : l \geqq 0]$ and $[lI : l \leqq 0]$.

Definition 4 *A finite globally causal orientation* in a Hausdorff topological space M is a transitive relation $x \prec y$ defined for suitable pairs $x, y \in M$, and called "precedence," having the properties that the set of all pairs (x, y) in $M \times M$ such that $x \prec y$ is closed, and that $x \prec x$ for all x. A *finite* (not necessarily globally) *causal orientation* in M is an assignment to each point $p \in M$ of a neighborhood N_p and a finite globally causal orientation in N_p such that if $q \in N_p$, then in a sufficiently small neighborhood of q, the given causal orientations in N_p and N_q coincide. If M is a C^∞ (respectively analytic) manifold, such a causal orientation is called C^∞ (respectively analytic) if it is locally definable by a finite set of C^∞ (respectively analytic) functions of x and y, in a fashion similar to that earlier indicated.

Example 3 Let $\mathbf{C}$ be a closed convex nontrivial cone in the real finite-dimensional linear space $\mathbf{L}$, and define $x \prec y$ to mean that $y \in x + \mathbf{C}$; $\mathbf{L}$ is thereby given a finite, globally causal orientation.

More generally, let G be an arbitrary topological (Hausdorff) group. A finite causal orientation in G is *right-invariant* in case $x \prec y$ if and only if $xa \prec ya$ for all $a \in G$; similarly for *left-invariant* and simply *invariant*. Let a *conoid* in G be defined as a closed subset K containing e such that $K^2 \subset K$. It is easily verified that for any conoid K, the relation $x \prec y$ if and only if $yx^{-1} \in K$ defines a finite globally causal right-invariant orientation in G; and that conversely every such orientation arises in this manner from the conoid $[x \in G : e \prec x]$; moreover such an orientation is fully invariant if and only if $aKa^{-1} \subset K$ for all $a \in G$.

More specifically, let $(\mathbf{L}, Q)$ be a *pseudo-Euclidean space*, defined as a pair consisting of a real linear space $\mathbf{L}$ of finite dimension together with a given nondegenerate symmetric bilinear form Q on $\mathbf{L}$; and suppose that Q is of type $(1, n)$, $n + 1$ being the dimension of $\mathbf{L}$, i.e., can be expressed in the form $x_0^2 - x_1^2 - \cdots - x_n^2$ in terms of suitable coordinates $x_0, x_1, \ldots, x_n$ on $\mathbf{L}$; such a pair $(\mathbf{L}, Q)$ will be called a *linear Lorentzian manifold*. The function $\varepsilon(X) = \operatorname{sgn} x_0$ defined on the closed subset $[X \in \mathbf{L}: Q(X, X) \geqq 0]$ is invariant under the component of the identity $O_0(\mathbf{L}, Q)$ of the automorphism group of $(\mathbf{L}, Q)$; and the subset in turn, $\mathbf{C} = [X \in \mathbf{L} : Q(X, X) \geqq 0, \varepsilon(X) \geqq 0]$, is a closed convex nontrivial algebraic cone in $\mathbf{L}$, which is invariant under $O_0(\mathbf{L}, Q)$. With the finite globally causal orientation determined by this cone, $(\mathbf{L}, Q)$ becomes the Minkowski space determined by $(\mathbf{L}, Q)$. It is also called the $(n + 1)$*-dimensional Minkowski space*.

The Minkowski spaces determined by two linear Lorentzian manifolds are isomorphic if and only if they have the same dimension. Here *isomorphism* means a one-to-one transformation preserving linearity, the form Q, and the precedence relation just defined; however, by the work of Alexandrov and Ovchinnikova, and Zeeman, cited earlier, the assumption of linearity is superfluous, if $n > 1$.

In case $n = 3$, the Minkowski space (of dimension 4) is isomorphic, in the same sense, to the space $\mathbf{H}(2)$ causally oriented by the cone $\mathbf{C}$ defined in Example 2. The isomorphism may be expressed in terms of the coordinates $t = x_0$, $x = x_1$, $y = x_2$, $z = x_3$, as follows:

$$X = (t, x, y, z) \to H = \begin{pmatrix} t - x & y + iz \\ y - iz & t + x \end{pmatrix}.$$

Since $X^2 = \det H$ and $2t = \operatorname{tr} H$, the causal cone in $\mathbf{L}$ (i.e. that defining in the indicated way the causal orientation, and definable as $[(t, x, y, z) : t^2 - x^2 - y^2 - z^2 \geqq 0,\ t \geqq 0]$) is mapped onto the set of all H such that $\det H \geqq 0$ and $\operatorname{tr} H \geqq 0$, i.e., the set $[H : H \geqq 0]$.

Definition 5 In a manifold M with an infinitesimal causal orientation, the set of all tangent vectors l at the point p such that $l \in C(p) \wedge -C(p)$ is called the *instantaneous present* at p, and denoted as $\mathbf{N}_p$. The causal orientation is called *Newtonian* if dim $\mathbf{N}_p$ (which is independent of p by continuity) is dim $M - 1$; *partially Newtonian* if dim $\mathbf{N}_p > 0$; *Einsteinian* if dim $\mathbf{N}_p = 0$; *Bergsonian* if dim $C(p) = 1$.

Similarly, in a manifold M with a finite globally causal orientation, the finite present P_x at a point x is defined as $[y \in M : x \prec y \text{ and } y \prec x]$. The orientation is called *Newtonian, partially Newtonian,* or *Einsteinian* near x_0 according as the equivalence classes relative to the relation $x \sim x'$ if $x \prec x'$ and $x' \prec x$ are totally ordered by the partial ordering canonically induced on them from that in M (for all x in some neighborhood of x_0) or $P_x \neq \{x\}$ (for some x in all sufficiently small neighborhoods) or $P_x = \{x\}$ (for all x in some neighborhood). It is *Bergsonian* if the union of the future and past of x is totally ordered. (We shall make little use of these definitions; they are included to suggest the conceptual basis and general scope of the theory.)

A (*causal*) isomorphism between causally oriented manifolds (of either the infinitesimal or finite type) is a manifold-isomorphism that carries the one causal orientation into the other. Similarly for the notion of (causal) automorphism of a causally oriented manifold; the group of all such, in the compact-open topology, will be called the (causal) *automorphism group* of the (causally structured) manifold. When the manifold is also a group, the term *causal morphism* may be used to avoid confusion with the notion of group automorphism.

Example 4 (a) Any left translation on a group with a left-invariant causal orientation (infinitesimal or finite) is a causal automorphism; but is not a group automorphism, except in the case of translation by the unit element. If the causal orientation is fully invariant, inner automorphisms are causal as well as group automorphisms.

(b) If in the first paragraph of Example 3 the cone $\mathbf{C}$ is proper, then according to a theorem of Alexandrov (1967), every causal automorphism is an affine transformation, whose homogeneous part leaves $\mathbf{C}$ invariant. In an arbitrary topological group G with invariant conoid K, any right or left translation is a causal-morphism, as is any group-automorphism of G which carries K into K; but in general, for a Lie group, the converse (i.e., the analogue of Alexandrov's theorem) is not valid.

(c) Note that the set of all "forward" vector fields (i.e., vector fields X such that $X_p \in C(p)$ for all p) is a convex cone in the space of all vector fields, M being taken here to be C^∞. Moreover, every tangent vector $l \in C(p)$ at some point p is the value at p of some forward vector field on M. Thus a causal orientation on a C^∞ manifold may equally well be defined by specification of the forward vector fields; and our axioms can be changed to the assumption that there is given a convex cone **C** in the space of all vector fields, which is closed in the topology of convergence on compact sets.

(d) Let S be an arbitrary C^∞ manifold, and set $M = R^1 \times S$. Define the vector field $a(\partial/\partial t) \times I_S + I_{R^1} \times X$, where X is any vector field on S, to be forward if and only if $a \geqq 0$. This defines a causal orientation on M which is evidently Newtonian. Note that the group of all causal automorphisms is infinite dimensional, for it includes all the transformations $I_{R^1} \times T$, for T an arbitrary diffeomorphism on S. (Here I_S denotes the identity operator on the space S.)

Scholium 2.1 A C^∞ manifold admits an infinitesimal causal orientation if and only if it admits a nonvanishing vector field.

If the C^∞ manifold M admits the nonvanishing vector field X, then defining $C(p)$ as $[aX_p : a \geqq 0]$ defines a causal orientation. To prove the "only if" part of the scholium, take a Riemannian metric on the manifold, thus obtaining in each tangent space a corresponding euclidean structure. Now note

Lemma 2.1.1 Let E denote a finite-dimensional Euclidean space, and K the set of all nontrivial closed convex cones in E. Then there exists a continuous Euclidean-invariant function defined on K, which maps each cone in K into a nonvanishing vector in the cone.

The topology on the space of cones is here the usual one, definable as that obtained from the Hausdorff metric applied to the intersections of the cones in question with the unit ball, the cones all being taken as having vertex at the origin. I am indebted to Professor W. Fenchel for the observation that this lemma is deducible from a result of Shepard (1966). To complete the proof of the scholium, assign to each point p the vector in $C(p)$ which is given by a map having the properties given in the conclusion of the lemma, the tangent space being taken as Euclidean in the indicated way. This shows the existence of a continuous nonvanishing vector field on M, whence a smooth such field also exists.

Remark 1 The variant of this result in which "causal orientation" is replaced by "Lorentzian structure" (i.e., hyperbolic pseudo-Riemannian metric) is well known; cf. Lichnerowicz (1971) for further developments in this direction. It is also well known that a compact manifold admits a nonvanishing vector field if and only if its Euler characteristic vanishes (Hopf–Samelson).

Example 5 Any covering manifold of a causally oriented manifold is naturally causally oriented itself, by virtue of the pullback from the local homeomorphisms into the covered manifold. It is easily seen that any covering manifold of a globally causal manifold is itself globally causal; but the covering manifold may be globally causal when the covered manifold is not, as e.g. in the case of $U(n)$, oriented as in Example 2. $U(1)$ is not globally causal, for the timelike arc $t \to e^{2\pi i t}$, $t \in [0, 1]$, is closed; but evidently the universal covering group $\tilde{U}(1) \cong R^1$ is globally causal, and the same is true for $\tilde{U}(n)$ (cf. below).

Definition 6 A *timelike arc* in a manifold with an infinitesimal causal orientation is a continuous, piecewise C^1, oriented arc whose forward tangent at each point p of the arc lies in $C(p)$; if in an extreme direction of the boundary of $C(p)$ for all p, the arc is called *lightlike*. Thus, if the arc is parametrized by the mapping $s \to p(s)$, $s \in [0, 1]$, then the tangent vector at $p(s_0)$,

$$f \to \lim_{\varepsilon \to 0} \varepsilon^{-1}[f(p(s_0 + \varepsilon)) - f(p(s_0))],$$

f being an arbitrary C^1 function near $p(s_0)$, is in $C(p)$, in the case of a timelike arc. The arc is *strictly timelike* if at each point the forward tangent lies in the interior of $C(p)$.

If p and q are points in the manifold M with infinitesimal causal orientation, we say "p precedes q" and write $p \prec q$ if there exists a timelike arc whose initial point is p and terminal point is q. It is evident that this relation is transitive: if $p \prec q$ and $q \prec r$, then $p \prec r$. If it has the property that $p \prec q$ and $q \prec p$ implies that $p = q$—alternatively, if every closed timelike curve is trivial, i.e. a single point—the causal orientation is said to be *semiglobal*. In such a manifold, the *future* F_x of any point x (respectively the *past* P_x or finite *present* N_x) is defined as the union of all points preceded by x (respectively which precede x, or both precede and are preceded by x).

Two points p and q in a manifold with infinitesimal causal orientation are called (relatively) *spacelike* if neither $p \prec q$ nor $q \prec p$. A *spacelike submanifold* is one any two points of which are relatively spacelike.

Example 6 (a) A compact manifold may admit a semiglobal causal orientation, an example being the n-dimensional torus, $n > 1$, oriented by taking **C** in its Lie algebra as displacement in the positive direction along an irrational one-parameter subgroup.

(b) On the other hand, *A compact C^∞ manifold with an infinitesimal causal orientation cannot be semiglobal if the causal cone at each point has nonvanishing interior*. For by the lemma cited earlier, there then exists a vector field X on the manifold M such that for each p, X_p is an interior point of $C(p)$. (Compare the proof of Scholium 2.1.) Let p_0 be a nonwandering point relative to the flow on M defined by X; by compactness, such a point exists. According to a version of the "closing lemma" due to Pugh,† there is a vector field X' on M which is arbitrarily close to X, (in the C^1 topology) and has a closed orbit. Such a vector field is however again timelike, in the sense that $X'(p)$ is interior to $C(p)$ for all p, i.e., M is not semiglobal.

(c) In Example 4c, the (finite) present at any point (t, q) consists of all points (t, q') with $q' \in S$.

(d) Any Einsteinian infinitesimal causal orientation whose cones $C(p)$ have nonvanishing interiors determines a finite causal orientation, by virtue of the relation $x \prec y$ earlier defined. This is a consequence basically of work of Zaremba and Marchaud as developed by Leray (1952); cf. also Choquet-Bruhat (1971). The concept of "global hyperbolicity" due to Leray and further developed by Choquet-Bruhat is a strengthening of the condition of global causality leading to global existence theorems of associated linear hyperbolic equations.

2. Causality in groups

A *causal group* is defined as a Lie group with an invariant causal orientation. Although in a vector group there are continuum many invariant causal orientations, in general Lie groups do not admit invariant causal orientations. We shall be particularly interested in cases in which they do admit such, but shall first discuss the general existence question.

Scholium 2.2 An open simple Lie group G admits an invariant causal orientation if and only if there exists an element $X \in \mathbf{G}$ such that if $a_1, \ldots, a_n$ are arbitrary in G and $c_1, \ldots, c_n$ are arbitrary nonnegative numbers, and if $\sum_j c_j \operatorname{ad}(a_j)(X) = 0$, then all $c_j = 0$.

Proof Note first that the instantaneous present of a simple group with invariant causal orientation necessarily vanishes. For it determines a linear subspace of the Lie algebra which is invariant under all inner automorphisms, and hence an ideal.

Now to show the "if" part of the scholium, define $\mathbf{C}$ as the closure of the set of all $\sum_j c_j \operatorname{ad}(a_j)(X)$, X being the fixed element of $\mathbf{G}$ which is given, and the a_j and c_j being as described in the scholium, and otherwise arbitrary. Then $\mathbf{C}$ is a closed convex invariant subset of the Lie algebra $\mathbf{G}$. It is nontrivial because if a convex set is dense in a finite-dimensional linear spacc, it must be all of the space. Thus if $\mathbf{C}$ is all of $\mathbf{G}$, every vector in $\mathbf{G}$ has the form $\sum_j c_j \operatorname{ad}(a_j)(X)$ for some nonnegative $c_1, \ldots, c_n$ and suitable $a_1, \ldots, a_n$ in G. But if $-X$ has this form, a contradiction to the hypothesis regarding X follows.

† Pugh (1967). It is possible to avoid the use of this result by a direct elementary argument due to L. Hörmander.

To prove the "only if" part of the scholium, let **C** be a nontrivial cone in **G** defining an invariant causal orientation, and let X be an arbitrary nonzero element of **C**. If $\sum_j c_j \operatorname{ad}(a_j)(X) = 0$ and not all $c_j = 0$, it follows that $-X = \sum_k c'_k \operatorname{ad}(a'_k)(X)$ for suitable nonnegative c'_k and elements a'_k of G, showing that $-X$ is in the instantaneous present. By the initial observation, this is in contradiction with the assumed simplicity of G.

Corollary 2.2.1 The group $O(n, 1)$ admits no invariant causal orientation if $n \geqq 3$.

Lemma Every element of the Lie algebra of $O(n, 1)$, $n \geqq 3$, is contained in the Lie algebra of some $O(3, 1)$ subgroup.

This follows by infinitesimalization of results of Wigner (1939) (cf. also Philips and Wigner, 1968).

Proof of corollary In view of the lemma, it suffices to show that every nonzero element of the Lie algebra of $O(3, 1)$ violates the condition of the scholium. Indeed, for every such element X there exists an element a of $O_0(3, 1)$ (where here and henceforth the subscript 0 to a group indicates the connected component containing e) such that $\operatorname{ad}(a)(X) = -X$. For as a Lie group, $O(3, 1)$ is locally isomorphic to $SL(2, C)$, and its Lie algebra correspondingly to that constituted by the 2×2 matrices of zero trace. Any such matrix is similar either to one of the form

$$\begin{pmatrix} l & 0 \\ 0 & -l \end{pmatrix} \quad \text{or to} \quad \begin{pmatrix} 0 & 1 \\ 0 & 0 \end{pmatrix}.$$

It is easily seen that any such matrix is similar to its negative. But a similarity transformation on such a matrix corresponds precisely to the action of $\operatorname{ad}(a)$ on the Lie algebra, for suitable a.

After this chapter was written, a general criterion for the case of semisimple groups was obtained by B. Kostant. With his permission, a slight modification of his treatment is given here.

Theorem (*Kostant*) Let G be a semisimple Lie group, and let R be a representation of G on the real finite-dimensional linear vector space **V**. Let K be a Lie subgroup of G such that $R(K)$ is maximal compact in $R(G)$. Then there exists an $R(G)$-invariant closed convex cone **C** in **V** such that $\mathbf{C} \cap -\mathbf{C} = \{0\}$ if and only if there exists a nonvanishing $R(K)$-invariant vector in **V**.

Proof It is evidently no essential loss of generality to assume that R is faithful, and to take R as the representation $A \to A$. Now suppose that **C** is a given cone with the indicated properties. Then there exists a linear functional λ on V such that $\lambda(x) \geqq 0$ for all $x \in \mathbf{C}$ and $\lambda(x_0) > 0$ for some $x_0 \in \mathbf{C}$.

Let $w = \int_{R(K)} Ax\, dA$; then w is K-invariant, and $\lambda(w) > 0$, showing that $w \neq 0$.

Conversely, suppose there exists a nonvanishing K-invariant vector w in $\mathbf{V}$. For the proof, a positive definite bilinear from β on $\mathbf{V}$ with the property that $\beta(Aw, w) \geqq 0$ for all $A \in G$ will be constructed. Toward this end, let $\mathbf{G} = \mathbf{K} + \mathbf{P}$ be the Cartan decomposition of $\mathbf{G}$, the Lie algebra of G, so that $G = PK$ is the polar decomposition of G, where $P = \exp \mathbf{P}$. Then $\mathbf{K} + i\mathbf{P} \equiv \mathbf{G}_u$ is a compact form of the complexification $\mathbf{G} + i\mathbf{G}$, and hence the complexification $\mathbf{V}_c = \mathbf{V} + i\mathbf{V}$ of $\mathbf{V}$ can be given a complex Hilbert space structure in such a way that the elements of $\mathbf{G}_u$ are skew-Hermitian. Let B denote the restriction to $\mathbf{V} \times \mathbf{V}$ of the Hilbert space inner product. Then all $X \in \mathbf{P}$ are Hermitian, which implies that A is positive definite for $A \in P$, i.e., $\langle Aw, w\rangle \geqq 0$ for all $A \in P$ and $w \in \mathbf{V}$. But any $B \in G$ is of the form $B = AU$, where $A \in P$ and $U \in K$. Since $Bw = Aw$, $\beta(Bw, w) \geqq 0$ for all $B \in G$.

Now let $\mathbf{C}_0$ denote the set of all finite linear combinations with positive coefficients of the Aw with $A \in G$. Then $\mathbf{C}_0$ is a G-invariant convex cone, and $\beta(v, w) \geqq 0$ for all $v \in \mathbf{C}_0$; it follows that $\beta(v, v') \geqq 0$ for all v and v' in $\mathbf{C}_0$. The closure $\mathbf{C}$ of $\mathbf{C}_0$ has the same properties. Finally, if z is both in $\mathbf{C}$ and $-\mathbf{C}$, then $\beta(z, z)$ and $\beta(z, -z)$ are both nonnegative, implying that $\beta(z, z) = 0$, and hence that $z = 0$.

Corollary If $\mathbf{G}$ is simple, then there exists a closed convex cone $\mathbf{C}$ in $\mathbf{G}$ such that $\mathbf{C} \cap -\mathbf{C} = 0$, and which is invariant under the adjoint representation, if and only if G/K is Hermitian symmetric, or equivalently, if the center of K has dimension 1.

Proof It is known that when $\mathbf{V}$ is irreducible, there exists at most one $R(K)$-invariant vector (within a scalar factor). The theorem then implies that the indicated cone $\mathbf{C}$ exists if and only if the dimension of the centralizer of $\mathbf{K}$ in $\mathbf{G}$ is 1. But according to a result of E. Cartan, as a real space $\mathbf{P}$ is irreducibly invariant under ad $\mathbf{K}$, and so contains no nonzero elements which commute with $\mathbf{K}$. It follows that $\mathbf{C}$ exists if and only if the center of $\mathbf{K}$ is one-dimensional, which, by another result of Cartan, is equivalent to G/K being Hermitian symmetric.

Discussion The maximal compact subgroup of $SO(p, q)$ is $SO(p) \times SO(q)$ implying that the Lie algebra of $SO(p, q)$ contains a cone $\mathbf{C}$ of the indicated type if and only if either p or q is two. On the other hand, it follows that $SU(p, q)$ (whose maximal compact subgroup is $SU(p) \times SU(q) \times U(1)$ for $pq \neq 0$), and $Sp(2n, R)$ (whose maximal compact subgroup is $U(n)$), and exceptional cases corresponding to E_6 and E_7 always admit such a cone.

In the case of a simple group, if there exists any nontrivial invariant convex cone $\mathbf{C}$, it is necessarily of the indicated type, since $\mathbf{C} \cap -\mathbf{C}$ is an ideal. In all probability, if $\mathbf{C}$ exists at all, it is unique (the integration argument given earlier shows it to be minimal), contains interior points, and coincides with the positive-energy cone for suitable unitary representations, i.e., the subset $\mathbf{C} = [X : iU(X) \geqq 0]$ for the representation U. In general, the conoid C in G generated by $\mathbf{C}$, i.e., the closure of $\bigcup_{n=1}(\exp \mathbf{C})^n$ in G lacks the important property that $C \cap C^{-1} = \{e\}$, i.e., G lacks a nontrivial *finite* sense of future displacement corresponding to the infinitesimal one defined by $\mathbf{C}$; but this is conjecturally the case if G is simply connected.

Scholium 2.3 In a causal group, the exponential map is locally causality-preserving (the Lie algebra being linearly causally oriented by its given cone): if $X \prec Y$, then $e^X \prec e^Y$ and if $e^{tX} \prec e^{tY}$ for all small $t > 0$, then $X \prec Y$.

Proof Since this is a local question, it is no essential loss of generality to take the group G to be a group of matrices (by Ado's theorem); the Lie algebra may then be identified with a Lie algebra of matrices. It is evidently sufficient (since dT is a linear isomorphism at each point in a sufficiently small neighborhood) to show that if T denotes the map $X \to e^X$, then dT carries any vector in the cone at X, $X + \mathbf{C}$ (identified by virtue of linearity with a subset of the tangent space at X) into a vector in the forward cone $C(e^X)$.

Consider the ray $[X + \varepsilon Z : \varepsilon \geqq 0]$, where Z is a fixed element of $\mathbf{G}$. This ray maps into the arc $e^{X+\varepsilon Z}$, $\varepsilon \geqq 0$, in G. To say that this arc is in a forward direction at $\varepsilon = 0$ is to say that $(\partial/\partial\varepsilon)e^{X+\varepsilon Z}e^{-X}|_{\varepsilon=0}$ lies in $\mathbf{C}$. In fact, by Duhamel's principle,

$$e^{X+\varepsilon Z} = e^X + \int_0^1 e^{(1-s)X}(\varepsilon Z)e^{s(X+\varepsilon Z)}\, ds.$$

From this it follows that $(\partial/\partial\varepsilon)e^{X+\varepsilon Z}e^{-X} = \int_0^1 e^{sX}Ze^{-sX}\, ds$. Since $\mathbf{C}$ is invariant and closed, the last integral has its value in $\mathbf{C}$.

To show that $e^{tX} \prec e^{tY}$ for all small t implies that $X \prec Y$; it suffices, noting that $e^{-tX}e^{tY} = e^{t(Y-X)} + O(t^2)$, to treat the case $X = 0$, which follows from the Leray (1952) theory.

Corollary 2.3.1 The unicover $\tilde{U}(n)$ of $U(n)$ (in the causal orientation earlier indicated) is globally causal.

Proof The unicover $\tilde{U}(n)$ is isomorphic to $R^1 \times SU(n)$, the covering transformation being $(t, u) \to e^{2\pi i t}u$. To show that $\tilde{U}(n)$ is globally causal it suffices to show that if (t_j, u_j), $j = 1, 2$, are any two points such that $(t_1, u_1) \prec (t_2, u_2)$, then $t_1 < t_2$, for there can then exist no nontrivial closed timelike arc. By compactness, it suffices to show this when the two points are arbitrarily close; they may then be taken to be in $U(n)$ rather than in the unicover, and by invariance it is no essential loss of generality to take one of the points as the group unit I. The result then reduces to showing that if w is sufficiently near e in $U(n)$ and if $e \prec w$, then $0 < t(w)$, where $t(w) = (2\pi ni)^{-1} \log \det w$; this is again a consequence of the Leray theory.

To show that the future and past of any point is closed, it suffices to use a criterion of Choquet-Bruhat (1971) for global hyperbolicity, according to which this is implied by the existence of a complete Riemannian metric on the manifold such that the timelike arcs from one point to another have bounded length. Using the direct product metric on $R^1 \times SU(n)$ (the usual one on R^1, any on $SU(n)$), this follows from the compactness of $SU(n)$ and what has been shown above.

Definition *7* A *forward displacement* in a causally oriented manifold M is a causal automorphism T such that $x \prec Tx$ for all $x \in M$.

Scholium 2.4 Let M be a manifold with infinitesimal causal orientation, whose corresponding finite relation $p \prec q$ defines a finite globally causal orientation (respectively, manifold with finite causal orientation). Let G be any C^1 Lie transformation group on M, which is represented by causal automorphisms of M. Let $\mathbf{C}$ denote the set of all elements X in $\mathbf{G}$ such that $\exp(tX)$ is a forward displacement for all $t > 0$ (respectively, near each point p is a forward displacement for all sufficiently small $t \geqq 0$). If $\mathbf{C} \neq 0$, then G is invariantly causally oriented by the designation of $\mathbf{C}$ as causal cone; and is globally causal if M is such.

Proof If $X, Y \in \mathbf{C}$, then $\exp[t(X + Y)] = \lim_n(\exp(tX/n)\exp(tY/n))^n$, which represents $\exp[t(X + Y)]$ as a limit of products of forward displacements; by the results just cited, any such product is again a forward displacement, as is any limit of such. Since $\mathbf{C}$ is invariant under multiplication by positive scalars by its definition, it follows that $\mathbf{C}$ is convex. Another application of the fact that the future of a point is a closed set shows that $\mathbf{C}$ is closed.

The elements of G act as causal automorphisms, and so transform by conjugation any forward displacement into another forward displacement. It follows that $\mathbf{C}$ is invariant under $\mathrm{ad}(G)$. Now if g is a function from $[0, 1]$ to G defining a timelike arc, and if x is arbitrary in M, then gx defines a timelike arc in M. If M is globally causal and if g is closed, it follows that gx is constant on $[0, 1]$, i.e., g is constant on $[0, 1]$, which means that G is globally causal.

Example *7* The causal automorphism group of Minkowski space M is the 11-parameter group consisting of the inhomogeneous Lorentz transformations augmented by scale transformations. This is a Lie group, which acts analytically, and so is globally causally oriented by the designation of $\mathbf{C}$ as those Lie algebra elements that generate forward displacements. This orientation is invariant, and consists of the sums of infinitesimal scale transformations with forward vector displacements.

Scholium 2.5 Let M be a C^∞ causal manifold admitting a connected Lie group G of causal automorphisms, of which M is a homogeneous space; and suppose that the subgroup H of G leaving fixed one point of M has finitely many components.

Then the universal covering space $\tilde{M}$ of M is $\tilde{G}/H'$, where H' is the connected subgroup of $\tilde{G}$ whose Lie algebra is (locally) the same as that of H, and $\tilde{G}$ acts causally on $\tilde{M}$.

Lemma 2.5.1 If G is a connected and simply connected Lie group and H is a closed connected subgroup, then $M = G/H$ is simply connected.

Proof Let $t \to m(t)$, $t \in [0, 1]$, be a continuous arc in M, with $m(0) = m(1) = e'$, where $e' = \phi(e)$, ϕ being the canonical map of G onto G/H; it must be shown that $m(\cdot)$ is homotopic to a trivial map. From the known local form of G/H, for any $s \in [0, 1]$ and for t sufficiently near to s, there exists a smooth arc $t \to g(t)$ in G such that $m(t) = \phi(g(t))$. Combining this with the simple connectivity of G, it follows that there exists a continuous arc $t \to g(t)$ defined for all $t \in [0, 1]$, such that $m(t) = \phi(g(t))$, $t \in [0, 1]$. Since the exponential map from the Lie algebra $\mathbf{G}$ to G has dense range, it is no essential loss of generality, for the purpose of showing that the arc $m(\cdot)$ is homotopically trivial, to assume that $g(1)$ lies on a one-parameter subgroup of H; otherwise, $m(1)$ and $g(1)$ may be displaced by arbitrarily little to achieve this situation, without affecting the homotopic character of the arcs in question. Now let $g'(t) = \exp(tX)$, where X is an element of the Lie algebra of $\mathbf{H}$ of H such that $\exp(X) = g(1)$. Then $g(\cdot)$ and $g'(\cdot)$ are homotopic in G; it follows that $\phi \circ g(\cdot)$ and $\phi \circ g'(\cdot)$ are homotopic in M; but the latter path is trivial.

Proof of scholium Let D be the discrete central subgroup of $\tilde{G}$ such that $\tilde{G}/D$ is isomorphic to G, and let θ denote the canonical homomorphism of $\tilde{G}$ onto G. Then $\theta(H') = H_0$, for $\theta(H')$ is a connected subgroup of G with the same Lie algebra as H_0, where H_0 is the component of the identity of H. It follows that the map $gH' \to \theta(g)H$ is well defined from $\tilde{G}/H'$ onto G/H_0. By general Lie theory and the fact that θ is a local isomorphism near the group unit, the indicated map is also a local homeomorphism, and hence is a covering transformation of $\tilde{G}/H'$ onto G/H_0, which in turn covers M finitely. Since $\tilde{G}/H'$ is simply connected by the lemma, it is the unicover of M.

In order to establish the causality of the action of $\tilde{G}$ on $\tilde{M}$, it suffices to show that every one-parameter subgroup of $\tilde{G}$ acts causally on $\tilde{M}$, for $\tilde{G}$ is generated by these, and any product of causality-preserving transformations is again such. Consider first the case in which $H = H_0$. Let ϕ denote the indicated covering transformation of $\tilde{M}$ onto M: $\phi(\tilde{g}H') = \theta(\tilde{g})H$. Then it follows that $\phi(\tilde{g}_1\tilde{g}_2H') = \theta(\tilde{g}_1)\phi(\tilde{g}_2H')$. Now observe the

Lemma 2.5.2 Define a diffeomorphism T at a point p of a causal manifold M as being *causal* at p in case dT_p carries C_p into C_{Tp}. Now let T_t be a one-parameter C^∞ group of diffeomorphisms of M, and suppose that for each point $p \in M$, there exists an $\varepsilon(p) > 0$ such that T_t is causal at p if $|t| < \varepsilon(p)$. Then T_t is causality-preserving, for every t.

Proof Let $\bar{t}$ denote the supremum of the values $t > 0$ such that T_s is causal at p for all $s \in [0, t]$; if $\bar{t} = \infty$, the conclusion of the lemma is valid, so suppose $\bar{t} < \infty$. Now T_s is causal at $T_{\bar{t}}p$ if $|s| < \varepsilon'$ for some $\varepsilon' > 0$. But this means that $T_{\bar{t}+s}$ is causal at p, contradicting the assumption that $\bar{t} < \infty$, and completing the proof.

To conclude the proof for the case $H = H_0$ it now suffices to show

Lemma 2.5.3 Every transformation on $\tilde{M}$ corresponding to an element of a one-parameter subgroup of $\tilde{G}$ is causality-preserving.

Proof Let $\tilde{p} = \tilde{g}_0H'$ be an arbitrary point of $\tilde{M}$, and let $\tilde{X}$ be arbitrary in $\tilde{\mathbf{G}}$. Then ϕ is locally a diffeomorphism near $\tilde{p}$, say ϕ is a diffeomorphism of the open set $\tilde{R}$ having $\tilde{p}$ in its interior onto the open subset R in M having $p = \phi(\tilde{p})$ in its interior. Specializing the relation indicated above, for any real t, $\phi(\exp(t\tilde{X})\tilde{g}_0H') = \exp(tX)\phi(\tilde{g}_0H')$, where $X = d\theta(\tilde{X})$. Let $\varepsilon > 0$ be so small that $\exp(t\tilde{x})\tilde{p} \in \tilde{R}$ and $\exp(tX)p \in R$ for $|t| < \varepsilon$. The two local one-parameter groups involved here then have equivalent action near $\tilde{p}$ and p, as implemented by ϕ; since $\exp(tX)$ is causality-preserving on M, this shows that $\exp(t\tilde{X})$ is causal at $\tilde{p}$, for sufficiently small t. It follows from the immediately preceding lemma that T_t is causality-preserving on $\tilde{M}$ for all t.

The general case reduces to the case in which $H = H_0$ once it is shown that the action of G on G/H_0 is causal; but the local action of G on G/H_0 is identical with its local action on M.

3. Causal morphisms of groups

We now consider groups of causality-preserving transformations on specific classes of causal groups. These transformations are not necessarily group automorphisms (e.g., vector translations on Minkowski space preserve causality but are not automorphisms of the vector group by which the space may be represented as a causal group); to avoid confusion, we shall use the term *causal morphism* rather than causal automorphisms to refer briefly to a causality-preserving analytic homeomorphism on a causal group.

The notation $\mathbf{H}(n)$ will refer to the real linear space of all $n \times n$ complex Hermitian matrices, as a causal linear manifold, the cone $\mathbf{C}(n)$ being taken as the matrices H which are $\geqq 0$. When indicated by the context, $\mathbf{H}(n)$ will be identified in the usual way with the Lie algebra of the $n \times n$ unitary group $U(n)$, whose causal orientation will be taken as that defined by the indicated cone.

Scholium 2.6 If $n \geqq 2$, every one-to-one transformation T of $\mathbf{H}(n)$ onto itself leaving 0 fixed and such that $T(H) \leqq T(H')$ if and only if $H \leqq H'$ is of the form $T(H) = G^*F(H)G$, where G is an arbitrary nonsingular matrix and F is either the map $F(H) = H$ or $F(H) = \bar{H}$.

Proof According to a theorem of Alexandrov (1967), any such transformation T is necessarily affine, in the real linear space $\mathbf{H}(n)$. The proof is concluded by reference to the result that any linear transformation on $\mathbf{H}(n)$ which is an isomorphism for the order relation has the indicated form.

Scholium 2.7 For any transformation $T = \left(\begin{smallmatrix} A & B \\ C & D \end{smallmatrix}\right)$ in $SU(n, n)$, let $\rho(T)$ denote the transformation $U \to (AU + B)(CU + D)^{-1}$ on $U(n)$. The map $T \to \rho(T)$ is a homomorphism of $SU(n, n)$ into the group of all causal morphisms of $U(n)$.

It is well known that the indicated transformation $\rho(T)$ does indeed act on $U(n)$, and that ρ is a homomorphism. To show that $\rho(T)$ is a causal morphism it suffices to show that for arbitrary $U \in U(n)$ and arbitrary Hermitian $H \geqq 0$, then

$$-i(\partial/\partial\varepsilon)T(e^{i\varepsilon H}U)T(U)^{-1}|_{\varepsilon=0} \geqq 0,$$

where $T(U)$ denotes $(AU + B)(CU + D)^{-1}$. By straightforward differentiation, the indicated derivative is

$$AHU(AU + B)^{-1} - (AU + B)(CU + D)^{-1}CHU(AU + B)^{-1}$$
$$= [A - (AU + B)(CU + D)^{-1}C]HU(AU + B)^{-1}.$$

To show that the last expression is nonnegative, it suffices to show that

$$A - (AU + B)(CU + D)^{-1} - C = [U(AU + B)^{-1}]^*.$$

This putative equality transforms by simple reversible operations into the putative equalities

$$A - (AU + B)(CU + D)^{-1}C = (U^*A^* + B^*)^{-1}U^*;$$
$$AU - (AU + B)(CU + D)^{-1}CU = (U^*A^* + B^*)^{-1};$$
$$(U^*A^* + B^*)^{-1}[(U^*A^* + B^*)AU - (U^*C^* + D^*)CU] = (U^*A^* + B^*)^{-1};$$
$$(U^*A^* + B^*)AU - (U^*C^* + D^*)CU = I,$$

and this last equation follows from the relations $A^*A - C^*C = I$, $B^*A - D^*C = 0$, which are implied by the assumption that $T \in SU(n, n)$.

Scholium 2.8 The Lie algebra of $U(n)$, as identified with $\mathbf{H}(n)$, is causally isomorphic to the open dense subset

$$U_*(n) = [U \in U(n) : \det(I - U) \neq 0],$$

via the Cayley transform $H \to (H - iI)(H + iI)^{-1}$.

Proof It suffices to show that if H and F are any fixed Hermitian matrices, and if $H(\varepsilon) = H + \varepsilon F$, then setting

$$U(H) = (H - iI)(H + iI)^{-1} \lim_{\varepsilon \to 0} (i\varepsilon)^{-1}[U(H(\varepsilon))U(H)^{-1} - I] \geqq 0$$

if and only if $F \geqq 0$, for the indicated limit relation means that the arc $\varepsilon \to U(H(\varepsilon))$ ($\varepsilon \geqq 0$) has a timelike forward direction at $\varepsilon = 0$. Now

$$\begin{aligned}(i\varepsilon)^{-1}[U(H(\varepsilon))U(H)^{-1} - I] &= -2(i\varepsilon)^{-1}[i(H + iI)^{-1} \\ &\quad - i(H(\varepsilon) + iI)^{-1}]U(H)^{-1} \\ &= 2\varepsilon^{-1}(H(\varepsilon) + iI)^{-1}[(H + iI) \\ &\quad - (H(\varepsilon) + iI)](H + iI)^{-1}U(H)^{-1} \\ &\quad - 2(H + iI)^{-1}F(H - iI)^{-1}.\end{aligned}$$

The last expression is of the form G^*FG where G is nonsingular, and so is nonnegative if and only if F is such.

Remark 2 It is very likely that all causal morphisms of $U(n)$ are of the form treated in Scholium 2.7. It is also likely that, within conjugacy, the Cayley transform is the most general open causal transformation from a linear causal manifold (i.e., one admitting a linear structure in such a way that the future of any point x has the form $x + C$ for some closed convex cone C) into $U(n)$, provided $n > 1$. (It is easily seen that this is not the case when $n = 1$.) This would follow directly if the Alexandrov theorem were true as a local theorem, but it is not: local conformal transformations in Minkowski space are local causal automorphisms, without necessarily being locally affine.

In a different but related vein, it is probable that any one-to-one transformation which preserves the causal structure in a smooth causal manifold is necessarily smooth, provided the defining causal cones are proper. (This means they should contain no full lines and should not be the direct product of a ray and a cone of lower dimension.) A partial result in this direction has been given by Choquet-Bruhat (1971); for manifolds which are globally hyperbolic in the sense of Leray, automorphisms in the indicated sense are necessarily continuous. It also seems probable that the group of all such automorphisms is finite-dimensional, again in the proper case. That some such restriction is necessary is shown by the case of two-dimensional Minkowski space; see Zeeman (1964).

Corollary 2.8.1 For any transformation $T \in \tilde{S}\tilde{U}(n, n)$, let $\tilde{\rho}(T)$ denote the action on the unicover $\tilde{U}(n)$ earlier indicated. Then $\tilde{\rho}$ is a homomorphism of $\tilde{S}\tilde{U}(n, n)$ into the group of all causal automorphisms of the globally causal space $\tilde{U}(n)$; and $\tilde{S}\tilde{U}(n, n)$ is itself globally causal with respect to the causal orientation naturally induced on it.

Proof This follows directly from earlier scholia, together with the results concerning $U(n)$ just obtained.

4. Causality and conformality

As already seen, Minkowski space is closely related to the space $U(2)$ in the series $U(n)$ of causal groups. It may also be represented in terms of projective quadrics in a way that brings out the relations between Minkowski spaces of different dimensions. Instead of the series $SU(n, n)$ of groups, the series $O(n, 2)$ intervenes. Considerations of conformality play a general role for this series.

Definition 8 A *conformal linear space* is a pair $(\mathbf{L}, \tilde{Q})$, where $(\mathbf{L}, Q)$ is a pseudo-Euclidean space (i.e., $\mathbf{L}$ is a real finite-dimensional linear vector space; Q is a nondegenerate symmetric bilinear form on $\mathbf{L}$), and $\tilde{Q}$ denotes the equivalence class of symmetric forms containing the given form Q, equivalence being defined as proportionality via a nonzero constant. For any given pseudo-Euclidean space $(\mathbf{L}, Q)$, the conformal linear space $(\mathbf{L}, \tilde{Q})$ is called the *induced* (or associated, or corresponding) conformal linear space; and $\tilde{Q}$ is called a linear conformal structure on $\mathbf{L}$.

A pseudo-Riemannian space is a pair (S, g) where S is a real C^∞ manifold and g assigns to each point x of S a nondegenerate symmetric bilinear form g_x on the tangent space S_x to S at x, in a C^∞ manner (i.e. in terms of local coordinates, the coefficients of g are C^∞). A conformal transformation from one pseudo-Riemannian space (S, g) into another (S', g') is a C^∞ homeomorphism T from S into S' such that for every point $x \in S$, the differential dT_x of T at x is a linear conformal isomorphism of $(S_x, \tilde{g}_x)$ into $(S'_{Tx}, \tilde{g}'_{Tx})$. When such a transformation T exists, the pseudo-Riemannian spaces (S, g) and (S', g') are said to be *conformally equivalent.*

A conformal space is a pair (S, q) consisting of a C^∞ manifold S together with a mapping q from each point x of S to a linear conformal structure on S_x, which near each point x_0 has the form $q_x = \tilde{g}_x$ for some pseudo-Riemannian structure g near x_0. Conformal transformations between conformal spaces, conformal equivalence of conformal spaces, etc. are defined correspondingly.

(The foregoing definitions are basically very well known; but because of slight variations in the literature, it has seemed desirable to make them explicit here; this serves also to indicate some notations.)

Now let $(\mathbf{L}, Q)$ be an arbitrary pseudo-Euclidean space. Let θ denote the mapping $x \to \tilde{x}$ from $\mathbf{L}$ onto the corresponding projective space $\tilde{\mathbf{L}}$ of all lines of $\mathbf{L}$. The manifold $\mathbf{Q} = [\tilde{x} \varepsilon \tilde{\mathbf{L}} : Q(x, x) = 0]$ is called the *projective quadric determined by Q.*

The orthogonal group $\mathbf{O}(Q)$ is defined to consist of all linear transformations T on $\mathbf{L}$ leaving invariant the form Q. If T is any nonsingular linear transformation on $\mathbf{L}$, the transformation $\tilde{T} : \tilde{x} \to \tilde{T}\tilde{x}$ is the projectivity induced by T. If $T \in \mathbf{O}(Q)$, then $\tilde{T}$ leaves M invariant, and the map $T \to \tilde{T} \mid M$ is a homomorphism of $\mathbf{O}(Q)$ into the group of all projectivities of $\tilde{\mathbf{L}}$ which leave M invariant. The image of $\mathbf{O}(Q)$ will be called the projective group on M, and denoted $\mathbf{P}(Q)$; locally, it is isomorphic with $\mathbf{O}(Q)$ via the indicated mapping.

Scholium 2.9 For any given real nondegenerate quadratic form on a finite-dimensional linear space $\mathbf{L}$, there exists a unique C^∞ conformal structure on the associated projective quadric which is invariant under the projective group of the quadric.

Specifically, this structure is given as follows in the notation just indicated:

Every tangent vector λ to $\mathbf{Q}$ at a point $\tilde{x}$ has the form $d\theta(\lambda')$ for some tangent vector λ' to $\mathbf{L}$ at x such that $Q(\lambda', x) = 0$ (making the canonical identification of the tangent space at x with $\mathbf{L}$), and conversely every vector of the form $d\theta(\lambda')$ is tangent to $\mathbf{Q}$ at $\tilde{x}$; and the linear conformal structure in the tangent space $T_{\tilde{x}}$ is determined by the quadratic form $g(\lambda_1, \lambda_2) = Q(\lambda'_1, \lambda'_2)$, where $\lambda_j = d\theta(\lambda'_j)$ $(j = 1, 2)$.

Proof That a unique conformal structure on $\mathbf{Q}$ is obtained in the indicated fashion is a matter of elementary calculus on manifolds, with the use of Euler's theorem on homogeneous functions. That this structure is invariant under the projective group follows from its invariant form. To show it is the unique such structure, it suffices to show that at each point $\tilde{x}$ of $\mathbf{Q}$, there exists a unique linear conformal structure in $T_{\tilde{x}}$ which is invariant under the induced action of the group of projectivities of $\mathbf{Q}$ which leave $\tilde{x}$ fixed. Taking the point at infinity (as is no essential restriction, the projective group being transitive on $\mathbf{Q}$), this is a matter of showing that the pseudo-Euclidean group on R^k, extended by magnifications, relative to a nondegenerate quadratic form Q' on R^k, leaves invariant no linear conformal structure other than that determined by Q'. This is elementary.

Definition 9 A real nondegenerate quadratic form is said to be of type (a, b) if it may be expressed in terms of suitable linear coordinates as $x_{-1}^2 + \cdots + x_{-a}^2 - x_1^2 - \cdots - x_b^2$. Types of pseudo-Riemannian and conformal structures are similarly defined.

Scholium 2.10 With the same notation as in the preceding scholium, if Q is of type (a, b), with $0 < a \leqq b$, then $\mathbf{Q}$ is analytically conformal with the direct product of two spheres $S^{a-1} \times S^{b-1}$ modulo the direct product of the corresponding antipodal maps.

Proof Taking $\mathbf{Q}$ to have the form given in the preceding definition, every point of $\mathbf{Q}$ is of the form $\tilde{x}$ with x such that

$$x_{-1}^2 + \cdots + x_{-a}^2 = x_1^2 + \cdots + x_b^2 = 1.$$

With $S^{a-1} = [(x_{-1}, \ldots, x_{-a}) : x_{-1}^2 + \cdots + x_{-a}^2 = 1]$, the mapping from $S^{a-1} \times S^{b-1}$ into $\mathbf{Q}$,

$$\Pi\colon x = [(x_{-1}, \ldots, x_{-a}), (x_1, \ldots, x_b)] \to \tilde{x},$$

is therefore onto M. Evidently, $\tilde{x} = \tilde{y}$ with x and y in $S^{a-1} \times S^{b-1}$ if and only if $x = \pm y$, so that Π is a twofold covering of $\mathbf{Q}$ by $S^{a-1} \times S^{b-1}$. The antipodal map $A\colon x \to -x$ is thus such that $S^{a-1} \times S^{b-1}$ modulo the two-element group $\{1, A\}$ is analytically isomorphic to $\mathbf{Q}$ via the indicated mapping.

Using the fact that a tangent vector to S^{a-1} at $(x_1, \ldots, x_a)$ has the form $\sum_{i=1}^{a} u_i(\partial/\partial x_i)$ with coefficients u_i such that $\sum_{i=1}^{a} u_i x_i = 0$, the u_i being unique, and that the length of this tangent vector according to the standard Riemannian structure on S^{a-1} is $\sum_{i=1}^{a} u_i^2$, it is straightforward to compute the conformal structure given by the preceding scholium in terms of the x_i as coordinates, and verify that it agrees locally with that on the direct product $S^{a-1} \times S^{b-1}$ (and so is the same as that at the quotient of this product modulo A).

Definition 10 A conformal space (S, q) is said to be *conformally* [globally] *causal* in case q is of type $(1, c)$ $(c \geqq 1)$, and if S admits a [global] causal orientation whose cone at any point x consists of one of the two cones in the tangent space at x on which $g(\lambda, \lambda) \geqq 0$, where g is any pseudo-Riemannian structure defined near x which induces the conformal structure q.

Remark 3 Any covering space of a conformal manifold is again a conformal manifold, in a unique way so that the defining covering local homeomorphism is locally conformal; and is conformally [globally] causal if the original manifold is such. Compare the earlier remark on the lifting of causal structures to covering manifolds.

Scholium 2.11 Let $(\mathbf{L}, Q)$ be a given pseudo-Euclidean space such that Q is of type $(2, n + 1)$, $n > 1$. Then the projective quadric $\mathbf{Q}$ defined above is conformally causal; and its unicover $\tilde{\mathbf{Q}}$ is globally so.

Proof Taking Q to be of the form $x_{-1}^2 + x_0^2 - \sum_{i=1}^{n+1} x_i^2$ relative to suitable coordinates x_i on **L**, and setting $\mathbf{Q}' = S^1 \times S^n$, then $\mathbf{Q}'$ has a natural conformal structure, i.e., the direct product of the structures on its factors, and as shown earlier, covers **Q** twice via a local homeomorphism which is also conformal. This conformal structure on $\mathbf{Q}'$ is also causal, as may be seen in the following way. The tangent space T_x at any point $x \in \mathbf{Q}'$ is a linear subspace of the set of all tangent vectors to R^{n+3}, where $\mathbf{Q}'$ is imbedded in R^{n+3} via the mapping

$$[(x_{-1}, x_0), (x_1, \ldots, x_{n+1})] \to (x_{-1}, x_0, \ldots, x_{n+1}).$$

This subspace consists of all tangent vectors to R^{n+3} of the form $\lambda = \sum_{i=-1}^{n+1} u_i(\partial/\partial x_i)$ such that $u_{-1}x_{-1} + u_0 x_0 = \sum_{i=1}^{n+1} u_i x_i = 0$. The conformal structure on $\mathbf{Q}'$ at x may be correspondingly determined by the pseudo-Riemannian structure g given by the equation

$$g(\lambda, \lambda) = u_{-1}^2 + u_0^2 - \sum_{i=1}^{n+1} u_i^2.$$

Now let $C'(x)$ denote the set of all vectors λ tangent to $\mathbf{Q}'$ at x, such that $g(\lambda, \lambda) \geqq 0$ and $u_{-1}x_0 - u_0 x_{-1} \geqq 0$. It is easily seen that $C'(x)$ is a closed convex cone, and that it is C^∞ as a function of x in the sense earlier indicated. Thus $\mathbf{Q}'$ is conformally causal. The conformal structure on $\mathbf{Q}'$ is invariant under A, and the same is true of its causal structure. For the tangent vector $\lambda = \sum_i u_i(\partial/\partial x_i)$ at any point $x \in \mathbf{Q}'$ is carried by (the induced action of) A into the tangent vector $\mu = -\sum_i u_i(\partial/\partial x_i)$ at $-x$. Evidently

$$g_{-x}(\mu, \mu) = g_x(\lambda, \lambda)$$

and

$$u_{-1}x_0 - u_0 x_{-1} = (-u_{-1})(-x_0) - (-u_0)(-x_{-1}),$$

showing that $\mu \in C'(-x)$ if and only if $\lambda \in C'(x)$.

The quotient manifold $\mathbf{Q} = \mathbf{Q}'/\{1, A\}$ therefore acquires both the conformal and causal structure of M, by taking $C(y)$ for $y \in M$ as $d\eta(C'(x))$, where η denotes the canonical map from $\mathbf{Q}'$ onto $\mathbf{Q}$. (Note that a conformal–causal structure is determined by its causal structure alone.)

The universal covering manifold $\tilde{M}$ is evidently $R^1 \times S^{b-1}$, the projection map from $\tilde{M}$ to Q' being

$$(t, (x_1, \ldots, x_b)) \to (\cos t, \sin t, x_1, \ldots, x_b).$$

To show that no timelike arc in $\tilde{\mathbf{Q}}$ is closed, let $s \to y(s)$, $s \in [0, 1]$ be an arbitrary such arc, and let $u(s)$ and $v(s)$ denote the components of $y(s)$ in R^1

and S^{b-1} respectively. It suffices to show that either $u(1) > u(0)$, or $\dot{y}(s) = y(0)$ for all s, for it then follows that the timelike arc is closed only if it is trivial. To this end it will suffice in turn to show that $u'(s) \geqq 0$, and $u'(s) = 0$ for all $s \in [0, 1]$ only if $y(s) = y(0)$ for all such s.

Observe in this connection that the transformations on $R^1 \times S^n$ of the form $T_1 \times T_2$, where T_1 is a translation in R^1 and T_2 is a rotation on S^n, are causal morphisms. In the case of T_1, translation through s is for sufficiently small s an action which locally is a causal morphism on $\mathbf{Q}$, i.e., its differential maps the defining cones $C(p)$ appropriately; it follows by continuity that this is true globally on $\tilde{M}$, for all s. In the case of T_2, the argument is similar. Thus in order to show that $u'(s) \geqq 0$ and that $u'(s) = 0$ only if $y'(s) = 0$, it suffices to consider the case in which $y(s) = 0 \times (1, 0, \ldots, 0)$. Further, since the question is a local one, it may equally be determined in $\mathbf{Q}'$ rather than $\tilde{\mathbf{Q}}$, with $y(s)$ taken as the point of $\mathbf{Q}'$ covered by $0 \times (1, 0, \ldots, 0)$, i.e., $(1, 0) \times (1, 0, \ldots, 0)$. (Note that the covering of $\mathbf{Q}$ by $\tilde{\mathbf{Q}}$ may be factored into the covering of $\mathbf{Q}'$ by $\tilde{\mathbf{Q}}$, followed by η.) Now writing

$$y(t) = (x_{-1}(t), x_0(t), \ldots,)$$

near $t = s$, where $y(s) = (1, 0, 1, 0, \ldots, 0)$, then $x_0(t) = \sin u(t)$, showing that $u'(t) = (\cos u(t))^{-1} x_0'(t)$ near this point. Observing that $\cos u(t) = x_{-1}(t) > 0$ near $s = t$, and that for a timelike arc from $y(s)$, $x_0'(t) \geqq 0$ by the requirement that $u_0 x_{-1} - u_{-1} x_0 \geqq 0$ for a tangent vector in the cone $C(y(t))$, it results that $u'(t) \geqq 0$. Further, $u(s) = 0$ only if $x_0'(s) = 0$; but then $x_{-1}'(s) = -\sin u(s) u'(s)$ showing that $x_{-1}'(s) = 0$; and the requirement that $u_{-1}^2 + u_0^2 \geqq \sum_{i=1}^n u_i^2$ for a tangent vector $\sum u_j(\partial/\partial x_j)$ in $C(y(s))$ then implies that $x_j'(s) = 0$ for all j, i.e., $y'(s) = 0$.

Corollary 2.11.1 $\tilde{\mathbf{Q}}$ is globally hyperbolic in the sense of Leray.

By a theorem of Choquet-Bruhat, it suffices to show that for every fixed pair of points, the set of all timelike arcs from one to the other is bounded relative to a complete Riemannian metric on $\tilde{M}$. The direct product of the usual (translation-invariant) metrics on R^1 and S^n is such. The argument just given shows that a timelike arc from (t_1, p_1) to (t_2, p_2) is such that the component is monotone increasing, while the S^n component describes an arc whose length over any interval is bounded by the length of the R^1 component. Thus the total length is bounded by $2|t_2 - t_1|$.

5. Relation to Minkowski space

We next show that the quadric $\mathbf{Q}$ just considered has imbedded in it (algebraically, conformally, and chronogeometrically) Minkowski space as an open dense subset. The imbedding transformation is an analogue to the Cayley transform† in its chronogeometric properties.

The concept of "lightlike point" has chronogeometric significance but we give here the following purely algebraic

Definition 11 Let $x, y \in \mathbf{L}$ be nonzero and such that $Q(x, x) = Q(y, y) = 0$. Then "$\tilde{y}$ is lightlike relative to $\tilde{x}$" means that $Q(x, y) = 0$.

In order to describe the cited imbedding quite explicitly it is helpful to recall some aspects of spherical geometry. A *sphere* in the pseudo-Euclidean space (M, F) is defined as a subset of the form $[X \in M : F(X - X_0, X - X_0) = k$, where X_0 and k are fixed in M and R^1, respectively; a *null-sphere* is one for which $k = 0$. Note that the equation defining a sphere can be put in the form

$$aF(X, X) - 2F(X, X_0) + c = 0,$$

where $a \neq 0$, and that conversely all such equations define spheres; null-spheres are characterized by the condition that $ac - F(X_0, X_0) = 0$.

A *conformal sphere* in M is a subset of the form $[X \in M : aF(X, X) - 2F(X, X_0) + c = 0]$, where a and c are fixed in R^1, X_0 is fixed in M, and not all of a, c, and X_0 vanish (in other words, a conformal sphere is either a sphere in the usual sense, relative to the Minkowski metric, or a hyperplane); a conformal null-sphere is one for which $ac - F(X_0, X_0) = 0$. Denoting as $\mathbf{L}$ the vector space of dimension $\dim M + 2$ whose components are a, the vector X_0, and c; and as Q the form

$$Q(a, c, X_0; a', c', X_0') = F(X_0, X_0') - (ac' + a'c)/2,$$

it follows that the set of all conformal null-spheres in M is in one-to-one correspondence with the projective quadric $\mathbf{Q}$. The canonical mapping from $\mathbf{L}$ onto the projective space of all its rays will be denoted as θ. X^2 will signify $F(X, X)$.

Scholium 2.12 The mapping j: $X \to \theta((1, X, X^2))$ from M into $\mathbf{Q}$ is conformal, and has range equal to the set of all points of $\mathbf{Q}$ that are not lightlike relative to the point $\theta((0, 0, 1))$.

Proof To say that $\theta((a, X, c))$ is lightlike with respect to $P_0 = \theta((0, 0, 1))$ is to say that $a = 0$. Thus $\theta((1, X, X^2))$ is never lightlike with respect to P_0.

† From an abstract standpoint the present treatment may in part be regarded as a chronogeometrical interpretation of the generalized Cayley transform known for symmetric spaces, applicable to certain Siegel domains.

Conversely, if $\theta((a, X, c))$ is not lightlike with respect to P, then it is no essential loss of generality to take $a = 1$, and then $c = X^2$.

To show conformality of the mapping j, note first that for any fixed vector A in M, the mapping $T: X \to X + A$, is conformal. Observe next that there exists a transformation T' on **Q**, in the conformal group treated earlier, such that $jT = T'j$. Indeed, T' is the transformation

$$(a, X, c) \to (a', X', c')$$

where $a' = a$, $X' = X + aA$, $c' = c + 2X \cdot A + aA^2$; this transformation is easily seen to be a projectivity which is in the group defined earlier, leaving **Q** invariant, for any value of A. Now since the totality of transformations of the form $X \to X + A$ is transitive on M, it follows that it is sufficient to show conformality at one point, say the point $X = 0$.

At this point, the differential of the mapping j: $X \to \theta(1, X, X^2))$ is the mapping dj: $Y \to d\theta((0, Y, 0))$, by a simple computation, with the usual identification of tangent vectors in M with vectors in M; and reference to an earlier scholium shows this to be conformal.

Scholium 2.13 A conformal transformation of a conformally causal manifold into a conformal manifold that admits a causal orientation is causal if the latter manifold is suitably oriented.

Proof Observe first that a conformal transformation from one conformal-causal manifold into another is either causal or anticausal (the latter meaning that the precedence relation is reversed). For if T denotes the transformation and $C^{\pm}(x)$ the infinitesimal future and past cones at x, then dT_x carries $C^{+}(x)$ into either $C^{+}(Tx)$ or $C^{-}(Tx)$. The set of all points x such that the former eventuality holds is open and closed by continuity, and the same is true of the latter eventuality; and "manifold" is always connected in the present usage.

Corollary 2.13.1 If F is of type $(1, n)$, then Q is of type $(2, n + 1)$, and j is causal if **Q** is suitably oriented causally.

Proof Choosing coordinates $x_0, x_1, \ldots, x_n$ such that $F(X, X) = x_0^2 - x_1^2 - \cdots - x_n^2$, and introducing variables x_{-1} and x_{n+1} by the equations $a = x_{-1} + x_{n+1}$, $c = x_{n+1} - x_{-1}$, then Q takes the form

$$Q(a, B, c; a, B, c) = x_{-1}^2 + x_0^2 - x_1^2 - \cdots - x_{n+1},$$

where $B = (x_0, \ldots, x_n)$. Thus **Q** admits a causal orientation, and j is causal by the preceding scholium, for a suitable choice of one of the two possible orientations.

Definition 12 When endowed with the conformal-causal orientation such that j is causal, $\mathbf{Q}$ will be called *n-dimensional conformal space–time*, or the *conformal compactification of n-dimensional Minkowski space–time.*

6. Observers and clocks

We now analyze mathematically the concept of observer, at increasing levels of specificity. In this connection one is naturally led to treat such concomitants of observers as "clocks" and "rods." The concepts developed coincide with mathematical forms of the usual physical notions in the case of Minkowski space; further examples are given in the cases of the two series of causal manifolds earlier considered.

Definition 13 Let M be a given globally causal manifold. A *spatio-temporal factorization* of M (for brevity, simply *factorization*) is an equivalence class of prefactorizations, where a prefactorization is a pair (S, ϕ) consisting of a C^∞ manifold S and a diffeomorphism ϕ of $T \times S$ onto M, where T is a real interval having the properties that:

(i) For any fixed $x \in S$, the map $t \to \phi(t, x)$ is a timelike arc in M;
(ii) For any fixed $t \in T$, the map $x \to \phi(t, x)$ defines a spacelike submanifold of M.

Two such prefactorizations (S, ϕ) and (S', ϕ') are *equivalent* if there exist diffeomorphisms f and g of R^1 onto R^1 and S onto S' such that f is orientation-preserving, and

$$\phi(f \times g)^{-1} = \phi'.$$

(Thus, corresponding to any factorization there are trivial fiberings of M by timelike arcs on the one hand, and by spacelike submanifolds (automatically maximal, as such), on the other. Conversely, two factorizations are the same if and only if the corresponding fiberings of M are the same.)

If a is any causal morphism of M, the *transform* of any prefactorization (S, ϕ) (respectively factorization represented by this prefactorization) is defined as the prefactorization (S, ϕ') (respectively factorization represented by this prefactorization), where $\phi'(t, x) = a(\phi(t, x))$; and the prefactorizations (respectively corresponding factorizations) are said to be *conjugate.*

Example 8 If M is $(n + 1)$-dimensional Minkowski space, and if $x_0, x_1, \ldots, x_n$ are linear coordinates such that the fundamental quadratic form $F(X, X) = x_0^2 - x_1^2 - \cdots - x_n^2$, a prefactorization may be defined by taking $S = R^n$ and defining $\phi(t, \mathbf{x})$ for arbitrary $t \in R^1$ and $\mathbf{x} \in R^n$ as the point of M having coordinates $(t, x_1, \ldots, x_n)$, where $\mathbf{x} = (x_1, \ldots, x_n)$. A scale transformation on M (i.e., similarity transformation), or a Euclidean transformation of the space component $(x_1, \ldots, x_n)$ (as a causal transformation on M) carries this prefactorization into a different one, which is, however, equivalent. Further, if $t \to t'$ and $\mathbf{x} \to \mathbf{x}'$ are diffeomorphisms of R^1 onto R^1 and R^n onto R^n, the former being orientation-preserving, then defining $\phi'(t, \mathbf{x}) = \phi(t', \mathbf{x}')$, the pair (R^n, ϕ') is equivalent to the pair (R^n, ϕ).

Definition 14 If S is a C^∞ manifold, and if R^1 and S have given Finsler structures (in the sense of norms in each tangent space, the norms being only positive-homogeneous), the *causal product* of R^1 and S (with the given structures) is the manifold $R^1 \times S$ with the causal structure which assigns to the point (t, x) the cone consisting of all tangent vectors $a(\partial/\partial t) + X$ (X being a tangent vector to S at x) for which $a \geqq 0$ and $\|X\|_x \leqq a\|\partial/\partial t\|_t$ (the subscripts indicating evaluation of the norms in the corresponding tangent spaces). A *metric prefactorization* of a causal manifold consists of a prefactorization (S, ϕ) together with given Finsler structures on R^1 and S such that ϕ is a causal morphism of $R^1 \times S$ onto M. A *metric observer* is an equivalence class of such, where (S', ϕ') with given Finsler structures on R^1 and S' is equivalent to the preceding one if and only if there exist maps f and g as earlier, with the additional property that f and g are Finslerian isometries.

Example 9 With the usual metrics on R^1 and R^n, the preceding factorization of Minkowski space is metric. As another example, consider the universal covering group $\tilde{U}(n)$ of $U(n)$, with the causal structure earlier indicated. The representation $\tilde{U}(n) \cong R^1 \times SU(n)$ determines a metric factorization, which is relative to the usual metric on R^1, and the following Finslerian metric on $SU(n)$ (which is Riemannian only for $n = 2$): with the identification of the Lie algebra of $SU(n)$ with $H_0(n)$ earlier indicated, $\|H\| = \inf[t : tI + H \geqq 0]$. It is easily verified that this defines a norm (positively, although not fully homogeneous, in general). Now if $H \geqq 0$ and $H = tI + H_0$ where H_0 is of zero trace, then necessarily $t \geqq 0$ and $\|H_0\| \leqq t$. Conversely, if $t \geqq 0$ and $\|H_0\| \leqq t$, then $tI + H_0 \geqq 0$ by the definition of $\|H_0\|$. Note that 0 is an interior point of the closed convex set in $H_0(n)$ consisting of elements of norm $\leqq 1$, although this set is not symmetric about the origin.

Definition 15 For any metric observer (S, ϕ) on a causal manifold M, the mapping from M into R^1 endowed with the given Finsler structure, defined by the equation $X \to t$ if $X = \phi(t, x)$ for some x, is called the *clock* of the observer and a clock on M is a mapping of the indicated type which is the clock of some observer. The mapping from M into S endowed with the given Finsler structure, defined by the equation $X \to x$ if $X = \phi(t, x)$ for some t, is called the *chart* of the observer; and a chart on M is a mapping from M to a C^∞ manifold endowed with a given Finsler structure, which is the chart of some observer. A chart and a clock are said to *match* if they are the chart and clock of one observer.

A causal morphism A is *forward* if $x \prec Ax$ for $x \in M$; *backward* if $Ax \prec x$ for all $x \in M$; *temporal* if either forward or backward; and *spatial* if x and Ax are relatively spacelike for all x. A group of causal morphisms of M is called temporal if it consists entirely of temporal transformations, and spatial if it consists entirely of spatial transformations.

A prefactorization (respectively factorization) is said to be *temporally homogeneous* if the map $\phi(t, x) \to \phi(t + t', x)$ is a temporal morphism, for all $t' \in R^1$; *spatially homogeneous* (respectively, and *isotropic*) if there exists a group G_s of causal morphisms of M (necessarily spatial) and an isomorphism ψ of G_s into the group of all diffeomorphisms of S, such that if $g \in G_s$, g: $\phi(t, x) \to \phi(t, \psi(g)x)$, and if the group $\psi(G_s)$ is transitive (respectively transitive and isotropic, or transitive on directions at any fixed point) on S; *homogeneous* if both temporally homogeneous and spatially homogeneous and isotropic; (respectively if a representative prefactorization is such). A *homogeneous observer* is an equivalence class of homogeneous prefactorizations, equivalence being defined as earlier. A metric and homogeneous observer whose prefactorization is homogeneous, and whose temporal and spatial groups (i.e., respectively the group $T_{t'}$: $\phi(t, x) \to \phi(t + t', x)$, or the group of all causal morphisms g such that for some diffeomorphism ψ of S, g: $\phi(t, x) \to \phi(t, \psi x)$, for all t and x) leave invariant the respective Finsler structures, is called *physical*.

A causal manifold is *covariant* if it admits a homogeneous factorization and in addition, the subgroup leaving one point p fixed is transitive on the strictly timelike directions (i.e., those in the interior of $C(p)$) at p. A *covariant observer* is a physical observer whose factorization is of this type.

If G_t is a given continuous one-parameter group of temporal transformations on M, with normalized parameter, a G_t-*clock* on M (where G_t is short for the transformation group (G_t, M)) is a function F from M to R^1 such that $F(T_t y) = F(y) + t$ for all $t \in R^1$ and $y \in M$. Similarly, if G_s is a given continuous group of spatial transformations on M, a G_s-*chart* consists of a C^∞ transformation group (G_s, S), the action of G_s on the C^∞ manifold S, being faithful, together with a map F from M to S such that $F(gx) = g(F(x))$.

Example 10 (a) *Minkowski space* A physical observer on this space M is defined by the earlier factorization, together with the unique invariant Riemannian metric on R^n invariant under the Euclidean group, which acts on M as a group of spatial transformations through its action on the component $\mathbf{x}$, and the unique translation-invariant metric on R^1, scaled so that causal cones $C(p)$ have the requisite form. Any two such observers, defined by coordinate systems of the indicated types, are conjugate within the causal morphism group. No other physical observers are known. As is well known, M is covariant, as are the indicated physical observers.

(b) *The spaces* $\tilde{U}(n)$ Taking $S = SU(n)$ and defining $\phi(t, u) = (t, u)$ for $t \in R^1$ and $u \in R^1$ and $u \in SU(n)$, then with the usual metric on R^1 and the Finsler metric on $SU(n)$ earlier indicated, and with the temporal group $T_{t'}$: $(t, u) \to (t + t', u)$; and spatial group $(t, u) \to (t, vuw)$ $(v, w \in SU(n))$, we have a physical observer. In addition to the Finsler metric on $SU(n)$ there is a unique invariant Riemannian metric on $SU(n)$, but this cannot in general be used to describe the causal structure on $\tilde{U}(n)$ in the way familiar in the case $n = 2$. $\tilde{U}(n)$ is covariant, for the causal morphism group is evidently transitive, so that it suffices to show that this group is appropriately transitive on the directions at a fixed point, say at the identity. This causal morphism group includes the action of $\widetilde{SU}(n, n)$ earlier indicated, which is locally identical to the action of $SU(n, n)$ on $U(n)$. Locally the Cayley transform is causal, so that this action can be transferred to the Lie algebra $\mathbf{H}(n)$, and then includes the transformations $H \to K^*HK + L$, where K is an arbitrary nonsingular matrix, and L is arbitrary in $H(n)$. Those transformations for which $L = 0$ leave 0 invariant, and the strictly timelike directions are those of a nonsingular $H \in H(n)$ such that $H > 0$. It is evident that if H' is another such direction, then there exists a nonsingular K such that $H' = K^*HK$. The spatial isotropy follows similarly and more readily.

(c) *The spaces* $\tilde{Q}_n$ This also is covariant and admits a physical observer; the Finsler metric on the space component is in this case Riemannian. With $S = S^n$ and $\phi(t, u) = (t, u)$ in our earlier notation, we have a prefactorization; with the usual metric on R^1 and the unique orthogonally invariant one on S, we have an observer. The temporal group $t \to t + t'$ acts appropriately to establish temporal homogeneity. The spatial group includes the action of the orthogonal group on S, lifted up to $\mathbf{Q}_n$, and this is evidently transitive and isotropic. Covariance follows from the facts that: (a) locally $\tilde{Q}_n$ is causally identical to Minkowski space; (b) the global causal morphisms of Minkowski space may all be lifted up to $\mathbf{Q}_n$; (c) the Lorentz group acts transitively on the strictly timelike directions at any point of Minkowski space.

(d) It should be noted that if F is any finite central subgroup of $SU(n)$, then $R^1 \times SU(n)/F$ is locally isomorphic to $\tilde{U}(n)$, and thereby defines a chronogeometry having *locally* all of its key symmetry properties; but that these properties may fail to be valid *globally*. For example, as pointed out by J. W. Milnor, if $n = 2$ and $F = \{\pm 1\}$, the corresponding factor space admits only a 7-dimensional causal morphism group, in contrast to the 15-dimensional group admitted by $U(2)$ itself. In particular, the factor space lacks global temporal isotropy. The proof reduces by general considerations to the determination of the causal vector fields on $U(2)$, which commute with F and thereby to a simple matrix computation.

Remark 4 (a) No other known (nonconjugate) covariant observers on the foregoing causal manifolds exist.

(b) It is likely that similar results hold for the Shilov boundaries of arbitrary Hermitian symmetric spaces, with suitable causal orientations; more precisely, for the unicovers of such manifolds. In all likelihood, the component of the identity of the causal morphism group is the induced action on the boundary of the group of the Hermitian space; and the chronogeometric features of the Cayley transform, which has already been extended to the general setting of such spaces, carry over.

7. Local observers

It is essentially straightforward to extend the foregoing considerations to local, rather than global, observers, using the usual concepts of the theory of local transformation groups. One may arrive in this way at a mathematical counterpart to the familiar physical concept of "local Lorentz frame."

If (S, ϕ) is a prefactorization on M, and if T and U are connected open subsets of R^1 and S, then with $\phi_1 = \phi \mid T \times U$, (U, ϕ_1) is an observer on $T \times U$, except that T is only diffeomorphic to R^1. We call this prefactorization on $\phi(T \times U)$ the restriction to this region of the given prefactorization on M. A local prefactorization at a point p in a manifold M is defined as prefactorization on some neighborhood N_p of p; two such are (locally) equivalent if their restrictions to some common neighborhood are (globally) equivalent; a local factorization at p is finally an equivalence class of such local prefactorizations.

Local observers may also be metric, homogeneous, physical, or covariant, the definitions being straightforward adaptations of the corresponding global ones. Conjugacy of local observers is defined as conjugacy via a local rather than global causal morphism, the causal morphism in question being one which preserves the relevant structure (metric, physical, or covariant).

Example 11 $\tilde{U}(n)$ is locally causally isomorphic to $\mathbf{H}(n)$, and Minkowski space M_n is locally causally isomorphic to $\tilde{Q}_n$. Hence, restricting to a suitable neighborhood the global observers on M_n and $H(n)$ previously given, and then transferring these observers via the aforementioned local causal isomorphisms to $\tilde{U}(n)$ and $\tilde{Q}_n$, one obtains certain local observers on these latter manifolds. In no case are these locally conjugate to the earlier given global observers, even as observers.

Remark 5 It is interesting to note that Haantjes (1937) has shown that any smooth local conformal transformation on a pseudo-Euclidean space can be extended to a global conformal transformation on its conformal compactification. In particular, any local smooth one-parameter group of local causality-preserving transformations on $\tilde{M}$ can be extended to a global such group acting globally on $\tilde{M}$. Conjecturally, this is valid for other cosmos associated with simple Lie groups, such as the series $SU(n, n)$, but this appears not to be known.

III

Physical theory

1. The Cosmos

Ultimately, any assertion about time, energy, and physical states must be transcribed into objective, experimentally verifiable statements, on whose validity the original assertion primarily depends for its own validation. However, it is often not possible, or desirable, to proceed in a purely logical-positivistic style, in which a physical theory is described solely in terms of predictions of the results of fully specified experiments. It is widely accepted that a general theoretical superstructure may be needed, or at any rate desirable, for a variety of reasons. Among these are economy and clarity of formulation, simplification of the means of correlation of the given physical theory with others, better adaptability to modifications which may prove desirable in other physical contexts, etc.

For these reasons, it seems desirable, and indeed perhaps necessary, to present our theory of the Cosmos from a viewpoint which is so fundamental and conceptually elementary that it may appear unfamiliar, and possibly overmeticulous. It seems especially important to approach the matter conservatively, because we attempt at the same time both to extend the direction in which special relativity departs from classical mechanics, and also to change the energy operator in quantum mechanics. It might appear desirable to separate these two developments, but they are logically very closely related, as will be seen later.

2. Postulational development

We now consider what may be deduced about the Cosmos on the basis of the following very general and broadly accepted assumptions.

Assumption 1 The Cosmos is a four-dimensional manifold.

Comment This means that in the vicinity of any point p of the Cosmos M, there is a four-dimensional coordinate system. It is of course a matter of the most elementary physical experience that in the vicinity of any observer, space–time events have a linear temporal order, and a three-dimensional position.

It might be objected that this assumption may eliminate singularities which could be significant from a general relativistic standpoint. The basic answer to this is that if these singularities be deleted, then the remaining region of space time is a regular manifold to which our considerations should then apply. Having settled the nature of this underlying regular manifold, one could then examine the adjunction of hypothetical singularities.

The second answer is that we wish to operate on as direct a level of experience as possible. Singularities in the space–time structure are theoretical possibilities of a definitely idealistic nature; their concrete analytical description involves delicate questions of the separation of physically essential aspects from matters of parametrization; until these have been materially clarified, it will be impossible to give operational meaning to an assertion as to the physical existence or nonexistence of space–time singularities.

Assumption 2 The Cosmos is endowed with a notion of causality.

Specifically: (a) at each point of the Cosmos there is given a convex cone of infinitesimal future directions, in the tangent space to the manifold at each point; (b) the future can never merge into the past, i.e., no curve that always points into the future can be closed.

Comment The existence of a sense of the infinitesimal future is a psychological fact; the physical meaning and implications of the notion of future are well developed in special relativity theory, and need not be repeated here. Bridgman has emphasized the logical independence of causality and the law for the addition of velocities in special relativity. This independence is indeed substantiated by the existence of space–time models that are globally acausal, while locally Minkowskian (cf. below). In such a model, the velocity of light would appear constant in all frames in the immediate vicinity of any observer, and the usual addition formula would hold, etc., but time would "wind back" on itself in the long run. This is counter to intuition, thermodynamics, and general physical ideas; while certain microscopic physical phenomena may well be cyclical in time, cyclicity of the Cosmos as a whole is generally implausible. We shall assume—without prejudice to future possibilities—that this is not the case.

The assumption of the convexity of the future cone in the tangent space at each point is not a matter of mere technical convenience, but is indicated by general conceptual considerations. One such consideration is that any displacement of the Cosmos which is the resultant of a succession of displacements *into the future* should itself be a displacement into the future. In particular, if X and Y are any two infinitesimal generators of one-parameter displacement groups into the future, in the sense that the group, denoted e^{tX}, generated by X carries each point p of the Cosmos M into a point $q = e^{tX}(p)$ which is temporally preceded by p when $t > 0$—in the sense that there is an arc from p to q whose forward tangent at every point is in the future direction—then $(e^{tX/n}e^{tY/n})^n$ should be a displacement of M into the future. But as $n \to \infty$, this displacement tends to $e^{t(X+Y)}$, a one-parameter group whose generator is $X + Y$. Thus, if X and Y are in the future direction, so also is $X + Y$, which means that the set of infinitesimal future directions at each point is closed under addition. Together with the evident fact that tX is always in the future direction if X is such and $t > 0$, this means that the set of all infinitesimal future directions at each point is convex.

Implicitly employed here is the extremely rudimentary assumption that a limit of points in the Cosmos which are preceded (or simultaneous with) a given point, is again such. This also means that the infinitesimal future at each point should be a *closed* set, in the mathematical sense. (For the most part in Chapter III we take for granted, unless otherwise indicated, such elementary points of mathematical regularity, and refer to Chapter II for formulations which are mathematically fully detailed.)

Assumption 3 The Cosmos admits stationary observers.

Comment It is difficult to see how any physical laws of the usual sort could be effectively discussed or verified without stationarity. For the dynamics, which form the crucial content of a complete physical theory, describe the change in state from one instant to another. Without a time-independent notion of state, such a description is evidently vacuous. Finally, in order to have a physical time-independent description of states, it seems necessary to have a stationary observer or equivalent operational means of labeling states.

Let us be quite explicit about the meaning of stationarity and of Assumption 3. We introduce the notions of "timelike" and "spacelike" in the fashion made possible by Assumptions 1 and 2, along customary physical lines.† Having done this, we can define a "forward displacement" as an admissible motion—i.e. a transformation of M onto itself which preserves causality, i.e., carries the totality of future directions at one point into those at the corresponding point—with the property that it carries each point into one that it (strictly) precedes (i.e., is the terminus of a strictly timelike arc originating at the original point). The concept of stationarity is then a relative notion; specifically, it is with respect to a one-parameter group of forward and backward displacements, or *temporal group*. The latter is defined as a family T_t of admissible motions of M, t being an arbitrary real parameter, such that

$$T_t T_{t'} = T_{t+t'} \qquad (t, t' \text{ arbitrary real numbers}),$$

$$T_t \quad \text{is a forward displacement for} \quad t > 0.$$

For example, in Minkowski space, if a denotes any fixed vector in the future cone, the family T_t: $x \to x + ta$ is of this nature. Conversely, in a Minkowski space of dimension greater than two, every temporal group has this form.

Thus, in order to have an effective notion of stationarity, it seems necessary that such an underlying temporal group of transformations be defined on the Cosmos. In general, however, a theoretical cosmos satisfying Assumptions 1 and 2 will admit no such group; indeed, in general there will be no admissible displacements. Assumption 3 thus carries first of all the implication that a temporal group exists. Beyond this, however, the usual notion of observer carries with it the implication that a space–time event is split by the observer into "space" and "time" components in a definite way. *An observer stationary with respect to the given temporal group* can be defined consonantly with this notion as one for whom this splitting into space and time components is unaffected by temporal evolution, as defined by the given group. More specifically, to each point p of the Cosmos, the observer assigns two components, t and x, where the time component t is a real number, while the space component x ranges over a three-dimensional manifold S. *A stationary observer*, with respect to the given temporal group, can now be defined as one such that the associated group T_t carries the point (t', x) of the Cosmos into the point $(t' + t, x)$ (more precisely, carries the point of the Cosmos with time and space components (t', x) according to his observation into the point with components $(t' + t, x)$). It can be shown that in Minkowski space this concept of stationary observer is equivalent to that of Lorentz frame, as one would expect. That is, there is a mutual correspondence between stationary observers and Lorentz frames, every such observer being associated with a Lorentz frame in the fashion earlier indicated.

† For further material on these notions and/or mathematical details, see Chapters I and II.

From an operational point of view, the "observer" is, in large part, in the present context simply this splitting of the Cosmos into time and space components; apart from such objective features, his existence is largely metaphysical. The splitting of space–time into space and time components relative to a complete local observational framework is partly a theoretical analysis and partly an empirical deduction from experience at a fundamentally more rudimentary level than a global theory of space–time. It is not merely a matter of anthropomorphic psychology, which as evidenced by the theory of relativity interacts nontrivially with theoretical ideas on the nature of space and time, but has a close relation to the concept of "stationary state" which is crucial in modern physics. Virtually all dynamics can be formulated as a description of transformations from one (at least approximately) stationary state to another; in particular, a temporal evolution group is required for an objective means of parametrization of states which can be correlated with experience. The parameters employed effectively define "space," particularly at extreme distances, the connection with the anthropomorphic notion of space being physically explicit only at the moderate-macroscopic level. In other words, "space" is defined by the condition that stationary state labels are (primarily) quantities (functions, vector fields, or operators) defined on space, together with the boundary condition that at middle distances, it coincides with the anthropomorphic notion. It is also limited by the conception that fundamental interactions are *local*, when expressed in spatial terms. The existence of degrees of freedom for elementary particles, which have not yet been correlated with geometrical space–time features (e.g., isotopic spin) does not essentially change these matters, since the only effect is to adjoin an "internal space" to space–time, which does not affect the physical space–time splitting.

The physically crucial notions *time* and *energy* are essentially immediate deductions from the formalism; they are defined relative to a given observer. The first component t in the space–time splitting is the observer's time; equivalently, it is the parameter of the one-parameter group T_t. The relation between the time of a space–time event, and time as the parameter of a temporal evolution group is simply that if $F(p)$ denotes the time of the space–time point p, relative to the given observer, for any t, $F(T_t p) = F(p) + t$, where T_t is the temporal evolution group associated with the observer. The temporal group invariance thus permits the correlation of time as an index of serial order with time as duration, an identification which is essential for real physics. At the same time, it uniquely specifies, apart from the choice of scale and zero point, the time parameter. The energy, on the other hand, is simply the conjugate or dual variable to time, i.e., the infinitesimal generator of temporal evolution. The situation thus is fundamentally more structured than in theories in which time appears primarily as an index of serial order—as, e.g., in general relativity as usually presented. This additional structure is of course not a technical burden, but rather an essential requirement for concrete physical interpretation. Without temporal invariance there is no conservation of energy—indeed, the very concept of energy becomes ambiguous; thus, despite intensive study, the precise formulation and properties of energy in general relativity appear to remain somewhat ambiguous.

Classically, Hamiltonian dynamics apply as readily to the space S as to the usual Euclidean configuration space. Quantum mechanically, the development likewise proceeds in entirely analogous fashion to the usual one. The dynamical variables are operators; the temporal evolution is defined, in, e.g., the case of a finite number of degrees of freedom, by a one-parameter group of unitary operators $U(t)$, the infinitesimal self-adjoint generator of which is the energy operator. This operator represents $-i(\partial/\partial t)$, which thereby defines the "energy" for the observer in question. Different observers will of course have different energy operators; depending on the geometry of their respective space–time splittings, these different energies may or may not be conjugate (in which case the eigenvalues are identical) or nonconjugate (in which case the eigenvalues are in general distinct; this theoretical possibility will be exemplified below).

Assumption 4 Space is homogeneous and isotropic.

It is entirely possible to conceive physical, and to give mathematical, examples of cosmos not satisfying this condition. However, it is intuitive, and substantiated at both macroscopic and microscopic levels. Moreover, as already suggested, it is physically essential to have some objective means of labeling particle states. The usual notions of angular and linear momenta, which have been found effective for this purpose, derive from the existence of just those symmetries which are here postulated. In all events, this postulate has been implicit in theoretical astronomy since the time of Cusanus.

Let us be quite explicit about what it means. Relative to any admissible observer, there is a splitting of space–time M into time and space components T and S; symbolically, $M = T \times S$, signifying that each point p in M corresponds to a pair (t, u), where the "time" t is in T, the range of time values (normally the real line), and the "spatial position" u is in S, the "space" of the observer. This splitting of space–time into one-dimensional time and three-dimensional space components is far from arbitrary; it is subject to the restrictions:

(a) For each fixed position in space u_0, the curve (t, u_0), where t varies over T, should be timelike. Indeed, this should be a maximal timelike curve, in the sense that any point which is timelike relative to (before or after) each point of the curve must already be on the curve.

(b) For each fixed time t_0, the submanifold (t_0, u), where u ranges over S, should be spacelike. Indeed, this should be a maximal spacelike submanifold, in the sense that any point which is spacelike relative to every point of the submanifold is already in the submanifold.†

These restrictions (a) and (b) are quite rudimentary and are totally independent of symmetry considerations. But as already noted, the correlations of the notion of observer with realistic physics leads naturally to the requirement of temporal homogeneity, without which one lacks a well-determined notion of energy. The requirement of spatial homogeneity is similar; it is not as fundamental as that of temporal homogeneity, but is tantamount to the intuitively plausible assumption that the laws of physics are independent of the physical location and orientation of axes. Philosophically speaking, it is undoubtedly possible to pursue physical theory without this assumption, but it would be extremely difficult to arrive at laws that were both nontrivial and definite.

Recently, difficulties in reconciling extragalactic astronomical observations with the expanding-universe model have led to proposals for limiting this postulate as regards the distribution of galaxies, if not for "empty" space itself. The work of G. de Vaucouleurs (1972) is representative of the observational background for such proposals, but can also (although not so construed by de Vaucouleurs) be interpreted as evidence against the expanding-universe model. In our view, the latter interpretation is more natural, and in fact, it will later be shown that the discrepancies studied by de Vaucouleurs may be resolved with a spatially homogeneous nonexpanding model (cf. also Sandage *et al.*, 1972).

In addition, the conventional theoretical microscopic picture—elementary particle analysis—is based on spatial homogeneity. The use of "linear momenta" as quantum numbers for particles is precisely tantamount to the assumption that spatial homogeneity is valid at the microscopic level.

For all these reasons, spatial homogeneity appears to be a quite reasonable postulate, in the simple form analogous to that of temporal homogeneity:

For any two points P and Q of "space" S, there exists a spatial transformation of the Cosmos, i.e., a smooth transformation $P \to P'$ of the cosmos M into itself, which preserves causality (i.e., carries relatively timelike and/or spacelike points into the same), and is spatial in the sense that any point P

† It would be just about as natural to define maximality in slightly different ways, e.g., in the spatial case as the absence of any spacelike submanifold of which the given one is a proper subset. However, all of the examples and applications of these notions treated here are maximal in both senses.

and its transform P' are relatively spacelike; and, for any time t, the point P corresponding to (t, p) is carried into the point P' corresponding to (t, p'), which carries P into Q.

Having dealt with spatial homogeneity, it is now a simple matter to deal with the similar notion of spatial isotropy, for which there are both ultra-macroscopic and microscopic forms of evidence at least as strong as those for spatial homogeneity. In mathematical terms, spatial isotropy means that given any two spatial directions at a point of the Cosmos, there exists a physically admissible transformation on the Cosmos (i.e., a causality-preserving smooth transformation) which carries one direction into the other. More precisely, if λ and λ' are any two tangent vectors at the point p of the space S, and if t is any time, then there exists a spatial transformation of the Cosmos which leaves the point P corresponding to the pair (t, p) fixed, and whose action on the tangent space at P carries λ into a nonzero multiple of λ'.

Applied to Minkowski space, these concepts naturally reproduce the usual ones. The only admissible (more specifically, "covariant") observers in the foregoing sense which are applicable to *all* of Minkowski space are obtained by a representation of the space in terms of pairs $(t, \mathbf{x})$, where t is a real number and $\mathbf{x}$ is a real vector, t being the temporal and $\mathbf{x}$ the spatial component, in the usual way. However, although this is the only *global* covariant observer, there are quite different *local* covariant observers. These cannot in general be extended to all of Minkowski space without encountering singularities; cf. below.

The concept of Lorentz frame is, in the case of Minkowski space, equivalent to that of observer in the present sense, except that the latter notion leaves unspecified a distance scale. Later, it will be shown how a specification of the distance scale may be accomplished on the basis of the present assumptions, without the presupposition of a given metric.

Example Let S be any three-dimensional Riemannian manifold, admitting an isotransitive group of isometries. (Here *isotransitive* means transitive both on points and directions at points; i.e., given any two points and directions at the points, there is a transformation in the group mapping the one point into the other, and the first given direction into the second.) Take as cosmos $M = R^1 \times S$, and define as the cone $C(t, q)$ at any point (t, q) of M the set of all tangent vectors of the form $a(\partial/\partial t) + \lambda$, where $a \geqq 0$, and λ is any tangent vector to S at q of length at most a. One then obtains an admissible cosmos, i.e., the foregoing assumptions are satisfied.

Spaces S of the indicated type have been completely classified by Tits (1957). Likewise classified are the four-dimensional Lorentzian manifolds (i.e., having a given pseudo-Riemannian structure whose fundamental form is of type (1, 3)) admitting certain types of transitivity, in an important work by Tits (1960). Physically speaking, the desiderata employed by Tits are of a qualitative relativistic nature. The physical process of observation, and the relation to symmetries defining the energy, etc., as considered here, supplement Tits' desiderata, and are materially restrictive. Thus, de Sitter space satisfies the cited qualitative relativistic desiderata, but admits no temporal translation group of the type earlier indicated (and thereby no natural definition of energy which results in a positive energy). Indeed, there are only three Lorentzian manifolds which satisfy Assumptions 1–5 on the Cosmos. The "universal space" $\tilde{M}$, consisting of the universal covering manifold of the conformal compactification $\bar{M}$ of Minkowski space M, has for its group of causality-preserving symmetries, one which is locally identical to $SO(2, 4)$ and so of dimension 15. Minkowski space can be regarded as an open dense submanifold of $\bar{M}$ which is covered infinitely often by $\tilde{M}$. Finally, the (twofold) covering space $S^1 \times S^3$ of $\bar{M}$, consisting of the direct product of a circle and the surface of a sphere in four-dimensional space, contains an open submanifold M' whose causal symmetry group corresponds to the subgroup $SO(2, 3)$ of $SO(2, 4)$. Thus, $\tilde{M}$ is universal also in the sense that the other two cosmos are simply derivable from it, and their causal symmetry groups are essentially subgroups of that of $\tilde{M}$. The cosmos represented by M' has only locally, not globally, the property that the region of influence of compact regions in space are compact; and the theoretical redshift–distance relation is unaffected if M' is used in place of M below; it thus appears as a slightly complicated variant of $\tilde{M}$ and will not be further considered here. It should perhaps be mentioned, however, that the causal structures in M and M' may be defined by metrics admitting ten-parameter isometry groups, while that in $\tilde{M}$ admits, at most, a seven-parameter group. Since Maxwell's equations are well-defined and invariant under the full causal symmetry group on all of the manifolds, the isometry groups of special metrics play no apparent physical role in the analysis of photon propagation; and the deviation from isometry in the case of $\tilde{M}$ is of order R^{-1}, where R is describable as the radius of the universe (cf. below) and so surely unobservable, even if physically meaningful in a local macroscopic theory.

The causal cones in the Lorentzian manifold case are defined by equations of second order. This is natural from the standpoint of general relativity, but there appears otherwise to be no inherent observational or physical reason why the causal cones should be of this special type. There exist simple models for which they are not quadratic, but satisfy all of the assumptions except that of four-dimensionality. The models $R^1 \times SU(n)$ discussed in Chapter II admit a quite satisfactory unique notion of causality, and Gårding (1947) has given an effective treatment of analogues to the Maxwell and Dirac equations in closely related spaces. Of course, the groups $SU(n)$ admit invariant Riemannian metrics, which thereby determine causal orientations on $R^1 \times SU(n)$ in the manner earlier indicated; however, this Lorentzian structure is much less invariant than the non-Lorentzian one (for $n > 2$), and in particular, unlike the latter, there is in general no symmetry in the theory which will transform one given timelike direction into another (cf. Assumptions 5 and 6). It is not yet determined whether any such nonquadratic models exist in dimension 4, but it appears unlikely.

To indicate how such models fit into the present scheme, it is appropriate to generalize the example just given by permitting the space S to be essentially Finslerian, rather than Riemannian. More specifically, we assume that there is given in the tangent space to S at each point q a closed convex body $K(q)$, containing 0 in its interior. Let $n(\lambda)$ denote the corresponding norm function for tangent vectors, i.e., $n(\lambda)$ is the largest nonnegative number s such that $s\lambda$ is in $K(q)$. One may then define the causal structure in M in the same way, except that $n(\lambda)$ is used in place of the length of λ. Thus, given any three-dimensional isotransitive Finsler manifold, there is a corresponding cosmos. *Conversely*, every admissible cosmos arises in this way from such a manifold, as indicated in Chapter II.

3. Physical observers

From the standpoint of rudimentary causality and homogeneity considerations, all of the foregoing models for the Cosmos thus appear equally good. We might now further refine these considerations and obtain additional plausible physical restrictions. For example, there are some further features of Minkowski space which might reasonably be postulated:

Assumption 5 Any given timelike direction at a point p is tangential to the forward direction of some admissible observer.

That is, the Cosmos M can be split into time and space, $M = R^1 \times S$, in such a way that p is represented by the point $(0, u_0)$, u_0 in S, and that the given timelike direction is represented by $\partial/\partial t$, t being the component in R^1. This assumption corresponds to the intuitive idea that there is no preferred direction in space–time, of temporal evolution, from which to observe the universe.

In a related vein, it would also be reasonable to postulate:

Assumption 6 Two different observers at the same point see the Cosmos in causally compatible ways, i.e., the transformation between their respective maps of the Cosmos should be causality-preserving.

It would be interesting to explore the consequences of these further assumptions, but we shall only remark here that both Minkowski space and universal space satisfy both of these assumptions. Rather than proceed in an increasingly abstract and somewhat philosophical line, it seems preferable to analyze in physical terms the process of measurement by which these models may be differentiated, and to which they are relevant. The space–time geometry itself is not necessarily directly observed; no apparent departures from a Euclidean model have been found by classical measurements. Rather, the geometry influences the analysis of microscopic (notably, elementary particle) and ultramacroscopic (notably, extragalactic) phenomena. It therefore seems physically more appropriate to correlate the geometry with what is observed in these extreme-distance realms. We shall begin with an analysis of the concept of "observer," and especially that of "local observer."

Since this may appear somewhat lengthy, we first summarize the salient points. Briefly, it will be found that:

(1) Minkowski and universal space are locally identical *as* causal manifolds.

(2) However, the natural clocks and physical observers in these two models are *not* equivalent; this means that there are two essentially distinct types of local clocks.

(3) If processes run and/or are observed by these inequivalent clocks, the lack of synchronization will be unobservable for times of the order of 1 yr, if the clocks are instantaneously synchronous. The asynchronization for times up to the order of 10^7 yr increases approximately quadratically with the time, and attains an observable level, in the form of the alteration it produces in the apparent frequency of a freely propagated photon. The relative shift $\Delta\nu/\nu$ is frequency independent.

One analytically simple means to represent the relation between Minkowski and universal spaces M and $\tilde{M}$, and particularly their admissible observers, is to utilize the well-known formulation of Minkowski space as the set of all 2×2 Hermitian matrices, and the relation of these matrices to the unitary 2×2 group, denoted $U(2)$. If Minkowski space M is coordinatized in the usual way by time and space coordinates (t, x, y, z), we may map M onto the space $\mathbf{H}(2)$ of all 2×2 complex Hermitian matrices by the transformation

$$F\colon (t, x, y, z) \to \begin{bmatrix} t + x & y + iz \\ y - iz & t - x \end{bmatrix}.$$

The crucial point here is that this mapping preserves causality, if a notion of temporal precedence is introduced into $\mathbf{H}(2)$ by the definition: H is "before" H' if $H' - H$ is a positive semidefinite matrix. (Any matrix, or point of Minkowski space, is considered to be both before and after itself, to simplify the terminology.)

Now the universal space $\tilde{M} = R^1 \times S^3$ can be usefully related to the special unitary group $SU(2)$ of all 2×2 unitary matrices of unit determinant in the following way. The most general matrix of $SU(2)$ has the form

$$\begin{pmatrix} a + ib & c + id \\ -c + id & a - ib \end{pmatrix},$$

with $a^2 + b^2 + c^2 + d^2 = 1$, i.e., with (a, b, c, d) on the three-dimensional sphere in 4-space defined by the equation $x_1^2 + x_2^2 + x_3^2 + x_4^2 = 1$. This is a one-to-one correspondence between S^3 and $SU(2)$, in which the unit matrix I corresponds to the point $(1, 0, 0, 0)$ on the sphere, and rotations of the sphere leaving this point fixed correspond to the transformations $U \to V^{-1}UV$, V being a fixed unitary matrix, on $SU(2)$. Note also that the usual invariant Riemannian metric on S^3 corresponds to the unique metric on the group $SU(2)$ which is invariant under both right and left translations on this group, i.e., the transformations $U \to VU$ and $U \to UV$, where V is a fixed element of the group $SU(2)$.

Locally, $\tilde{M} = R^1 \times S^3$ may be made to correspond to the 2×2 unitary group $U(2)$ by the mapping $(t, p) \to e^{it}u$, where $p \to u$ is the mapping from S^3 into $SU(2)$ just indicated. Sufficiently near to the point $t = 0, p = (1, 0, 0, 0)$, this mapping is one-to-one and smooth. A crucial point is that it is also causality-preserving, if a local notion of temporal precedence is introduced into $U(2)$ by taking as the set of future directions at the unit matrix I, all those that are represented by positive semidefinite Hermitian matrices; and at any other point defining the future directions by translating in the group $U(2)$ from I to the point in question. (Because of the invariance of the set of positive semidefinite matrices under the transformations $H \to UHU^{-1}$, U unitary, it is immaterial whether right or left translations are used.)

To set up a local causality-preserving transformation between Minkowski and universal space, it therefore suffices to set up such a mapping between $\mathbf{H}(2)$ and $U(2)$, which carries $H = 0$ into $U = I$. The simplest is the Cayley transform:

$$\Lambda: H \to \frac{2I + iH}{2I - iH}.$$

(For the proof that this is causality-preserving, see Chapter II; here 2 may be replaced by an arbitrary nonvanishing constant, but the present normalization will be convenient later.) These mappings are also conformal, in as much as a causality-preserving transformation on a pseudo-Riemannian causal manifold whose future cone is defined by the given metric is always conformal.

Consider how a given Lorentz transformation in Minkowski space M

appears from the standpoint of a locally equivalent observer on unispace $\tilde{M}$. Temporal evolution in M is

$$H \to H + sI.$$

Under the correspondence Λ, this transforms into a complex mapping in $\tilde{M}$ which mixes up the space and time components. The same is true of the space translations in a fixed direction,

$$H \to H + sK, \qquad K \text{ fixed in } \mathbf{H}(2), \quad \operatorname{tr} K = 0.$$

Thus time and space displacement in Minkowski space do not at all correspond via the causal mapping Λ to natural time and space displacements in $\tilde{M}$.

The natural space displacements in $\tilde{M}$ are given by the six-parameter group of rotations of the sphere S^3. Equivalently, this group consists of left and right translations on $SU(2)$, i.e., the transformations $V \to WVW'$, where V ranges over $SU(2)$ and W and W' are arbitrary fixed elements of $SU(2)$. Only the rotations, say those leaving the identity matrix I fixed, which are the transformations $V \to U^*VU$, where U is a fixed element of $SU(2)$, correspond precisely to conventional spatial displacements, i.e., the normal space rotations about the origin in R^3.

It is indeed impossible to set up between these spaces a conformal equivalence that does not mix space and time components in one space or the other. Actually, the space $\tilde{M}$ occurs in this discussion primarily as a means of exposing the central fact that Minkowski space admits two conceptually equally valid types of local physical observers, which are essentially distinct. Each observer sees space–time as split into space and time components in accordance with the underlying causal structure; each of them admits local temporal and spatial symmetries, acting on the time and space components separately, and being tantamount to the usual hypotheses of conservation of energy, linear momentum, and angular momentum; each has a unique notion of temporal duration and spatial distance (within scaling); each admits transformation to accelerated observers (i.e., there is no preferred strictly timelike direction). As far as general physical considerations go, there is no significant basis to prefer the one type of local observer to the other.

It might be argued that the conventional splitting into space and time components is "simpler"; it is "flat," while the other is "curved" (as regards space); it is traditional. On the other hand, there is no direct observational basis for asserting that the Cosmos is Minkowskian for very large times and distances. If indeed it should conform to the global extension of the unconventional splitting, the latter splitting would have the advantage of being much more symmetrical than the conventional Minkowskian one; it admits a 15-parameter group of admissible transformations on local frames of reference, while in the case of Minkowski space this group is only 11-dimensional. It will be seen that M is in a natural way, associated with any given observer, essentially contained in $\tilde{M}$, denumerably often, and one could argue thereby that $\tilde{M}$ may be more appropriate for the description of very long-range phenomena.

However this may be, it appears that a conclusive physical preference for one type of observer over the other can only be based on quantitative observation. A priori, both types could be valid, in the sense that different phenomena require different types of observers for their simple description. Indeed, it is instructive to compute how the frequency as measured by the one type of observer appears relative to one of the other type, so chosen as to be locally (more precisely, infinitesimally) at rest relative to the first one.

In order to make this computation, it is convenient to use the formalism of conformal space developed by Veblen and others. This formalism is fundamental in the present theory and we pause in our physical discussion to develop simple aspects of it in a somewhat rounded way.

4. Conformal geometry and the unitary formalism

It will be just as easy to take an $(n + 1)$-dimensional Minkowski space M with coordinates $t, x_1, \ldots, x_n$ (in fact, the case $n = 1$ will be relevant and illuminating). Our first step is to define a closed (compact) space $\bar{M}$, the so-called conformal space, in which M is imbedded naturally. Roughly speaking, $\bar{M}$ is obtained from M by adding a light cone at infinity. This does not disturb the underlying symmetry of the space; the Lorentz group and scale transformations continue to act conformally on $\bar{M}$. Indeed, one gains in symmetry, in that conformal inversion is a nonsingular operation on $\bar{M}$, and together with the Lorentz and scale transformations, generates a 15-dimensional Lie group which acts in an entirely regular and conformal manner on $\bar{M}$. (This is only a preliminary step; the space $\bar{M}$ is acausal; however, the remedy will be conceptually simple, and $\bar{M}$ will remain as a fundamental object for many computations.)

Specifically, $\bar{M}$ is the space of all (projective) conformal spheres in M of zero radius, endowed with the natural Lorentzian structure (cf. Chapter II). Analytically, let

$$X = (t, x_1, \ldots, x_n), \qquad t = x_0\,, \qquad X^2 = t^2 - \mathbf{x}^2$$

$$\xi_{-1} = 1 - \tfrac{1}{4}X^2, \qquad \xi_{n+1} = 1 + \tfrac{1}{4}X^2, \qquad \xi_j = x_j\,.$$

Set $\Xi = (\xi_{-1}, \xi_0, \ldots, \xi_n, \xi_{n+1})$, $\Xi^2 = \xi_{-1}^2 + \xi_0^2 - \xi_1^2 - \cdots - \xi_n^2 - \xi_{n+1}^2$ for arbitrary ξ_j; let Λ denote the transformation $\Lambda\colon X \to \Xi$ (given in terms of X

by the foregoing equations); then $\Xi^2 = 0$ if $\Xi = \Lambda(X)$ for some X. Set $\boldsymbol{\varepsilon} = (\varepsilon_{-1}, \varepsilon_0, \ldots, \varepsilon_{n+1}) = (1, 1, -1, \ldots, -1)$, so that $\Xi^2 = \sum \varepsilon_i \xi_i^2$. Let W denote the $(n+3)$-dimensional vector space of all Ξ (with arbitrary real values for the ξ_j). Let $\tilde{W}$ denote the $(n+2)$-dimensional projective space of rays in W (i.e., vectors Ξ when proportional vectors are identified). Let Q denote the quadric in W defined by the equation $\Xi^2 = 0$, and let $\tilde{Q}$ denote the corresponding quadric in $\tilde{W}$. Let $\tilde{\lambda}$ denote the map

$$\tilde{\lambda}: X \to \tilde{\Xi},$$

where $\tilde{\Xi}$ denotes the ray through Ξ. Then our earlier observation is to the effect:

$$\tilde{\lambda} \quad \text{is a one-to-one mapping of} \quad M \text{ into } \tilde{Q}.$$

More specifically, one can recover the $t, x_1, \ldots, x_n$ from a point of $\tilde{Q}$ (other than the exceptional points which do not correspond to points of M) as follows. For any nonzero Ξ in W, let $\mathbf{u}$ denote the vector $(u_{-1}, u_0, u_1, \ldots, u_{n+1})$, with $u_j = \xi_j/(\xi_{-1}^2 + \xi_0^2)^{1/2}$; we set $\mathbf{u} = u(\Xi)$, and note that u is the same for all vectors on the same ray as Ξ; the definition $\tilde{u}(\tilde{\Xi}) = u(\Xi)$ is therefore unique, and $\tilde{\Xi} \to \tilde{u}$ is a well-defined mapping from Q into an $(n+2)$-dimensional space R^{n+2}. It is clear from the definition that

$$u_{-1}^2 + u_0^2 = 1 = u_1^2 + \cdots + u_{n+1}^2.$$

Thus this mapping from $\tilde{Q}$ into R^{n+2}, to be denoted Γ, actually maps $\tilde{Q}$ into the direct product of a circle S^1, coordinatized by u_{-1} and u_0, and a sphere S^n, coordinatized by $u_1, \ldots, u_{n+1}$. Conversely, any point of this direct product $S^1 \times S^n$, $\mathbf{u} = (u_{-1}, \ldots, u_{n+1})$, corresponds to a point of $\tilde{Q}$ via the mapping $\gamma: \mathbf{u} \to \tilde{\Xi}$, where $\Xi = (u_{-1}, \ldots, u_{n+1})$. The mapping γ is precisely two-to-one, for both $\pm\mathbf{u}$ correspond to the same point of $\tilde{Q}$; we say γ is a *twofold covering* of $\tilde{Q}$ by $S^1 \times S^n$.

Now suppose one is given the coordinates $(u_{-1}, \ldots, u_{n+1})$ of one of the two points of $S^1 \times S^n$ corresponding to a given point X in M. Then the u_j are given by the equations

$$u_{-1} = k^{-1}(1 - \tfrac{1}{4}X^2), \qquad u_{n+1} = k^{-1}(1 + \tfrac{1}{4}X^2),$$
$$u_j = k^{-1}x_j, \qquad k = \pm[(1 - \tfrac{1}{4}X^2)^2 + t^2]^{1/2}.$$

The mapping $X \to \mathbf{u}$ is one-to-two, but is locally one-to-one. In the vicinity of the point $X = 0$, then, with k chosen to be positive (so that the point corresponding to $X = 0$ is $\mathbf{u} = (1, 0; 0, \ldots, 0, 1)$, we may recover X by the equations

$$x_j = 2u_j(u_{-1} + u_{n+1})^{-1} = 2\xi_j(\xi_{-1} + \xi_{n+1})^{-1}.$$

Now $S^1 \times S^3$ has a natural Lorentzian structure, defined by the form $du_{-1}^2 + du_0^2 - du_1^2 - du_2^2 - \cdots - du_{n+1}^2$; or, introducing the new time $\tau = \tan^{-1} u_0/u_{-1}$ near the point indicated, $d\tau^2 - ds^2$, where ds denotes the usual element of length on the sphere S^n. This Lorentzian structure is not invariant under the action of the Lorentz group in M, after transference to $S^1 \times S^n$ by the foregoing correspondences; however, it remains conformally invariant. More generally, any of the classical conformal transformations on M (which apart from the Lorentz transformations and the scale transformations are not everywhere defined on M, and develop singularities if an attempt is made to extend their domains of definition in M) act without singularities on $S^1 \times S^3$ and also on Q; and are conformal on these spaces, in each of which M is in a certain sense contained.

Moreover, the space $S^1 \times S^3$ admits a local notion of causality, according to which the one of the two convex cones defined by the Lorentzian form $d\tau^2 - ds^2$, which is in the direction of increasing τ, defines the future direction in space–time. Of course, it is always possible in a Lorentzian manifold to introduce a local notion of causality by a choice of one of the two cones near a point as the future; but it is not always possible, as it is here, to do so in a manner which varies continuously from point to point, throughout the global space–time manifold. Despite the latter feature, however, $S^1 \times S^3$ is not causal in the large; indeed, the curve

$$u_{-1} = \cos\tau, \qquad u_0 = \sin\tau, \qquad u_i = c_i \quad (\text{const}), \qquad 1 \leqq i \leqq n+1,$$

is in the forward timelike direction (as a function of τ), but is cyclic.

Thus, locally at any point in Minkowski space, we have the two distinct type of coordinates: t, x, y, z on the one hand, and the spherical coordinates u_j $(-1 \leqq j \leqq 4)$ on the other. There are many other parametrizations, of course; what is distinctive about the foregoing ones (or variants thereof) is their extensive covariance. To make this explicit, consider the actions of fundamental symmetries in terms of these parametrizations.

On the $(n+3)$-dimensional vector space Ξ, let L'_{ij} denote the vector field

$$L'_{ij} = \varepsilon_i \xi_i D_j - \varepsilon_j \xi_j D_i, \qquad \text{where} \quad D_i = \partial/\partial\xi_i .$$

These generate the group $O(n+1, 2)$ of linear transformations in Ξ leaving invariant the quadratic form Ξ^2. Consequently, they determine corresponding vector fields L_{ij} on the projective quadric $\tilde{Q}$. As shown in Chapter II, the conformal structure on $\tilde{Q}$ is invariant under the transformations generated by the L_{ij}, i.e., they are all infinitesimally conformal. Conversely, all globally defined infinitesimally conformal transformations on $\tilde{Q}$ are linear combinations of the L_{ij}. It is important to note that the full group of symmetries of Minkowski space, including all Lorentz transformations and changes of scale, extend from the space M to the larger space $\tilde{Q}$, and to its twofold covering $S^1 \times S^3$, remaining all the while causality-preserving (in particular, conformal); and no singularities are involved in this extended action.

It is useful and of interest to calculate the form of the generators L_{ij} when expressed in terms of the usual Minkowski coordinates; and on the other hand, to calculate the form of the Lorentz transformations in terms of the ξ_i or spherical coordinates on $S^1 \times S^3$. In addition, the natural symmetries of $S^1 \times S^3$ relative to the given splitting (i.e., transformations acting on S^1 alone, respectively acting on S^3 alone) will be of interest. These matters are more readily achieved and better understood if further parametrizations of space–time closely related to the given ones are developed. We shall therefore treat parametrizations in terms of $U(2)$ and the space of conformal null spheres in Minkowski space.

In Minkowski space M, a *Lorentz sphere* is a locus in M consisting of all vectors X such that

$$(X - X_0)^2 = \text{const} \qquad (X_0 \text{ a fixed vector});$$

if the constant = 0, one has a *null sphere*. Now a Lorentz sphere has an equation of the form

$$aX^2 - 2B \cdot X + c = 0$$

with $a \neq 0$ and B and c arbitrary; this sphere is a null sphere if and only if $ac - B^2 = 0$ (here B is a vector in Minkowski space). Evidently, there is a one-to-one correspondence between points of M and null spheres (a, B, c), where (a, B, c) is considered as a point of projective $(n + 2)$-dimensional space, and (a, B, c) is restricted to lie on the quadric $ac - B^2 = 0$. (More specifically, (a, B, c) and (a', B', c') are identified if there exists a nonzero constant λ such that $\lambda(a, B, c) = (a', B', c')$.)

In addition to the normal null spheres just indicated, there exist "ideal" null spheres given in the same way except that $a = 0$. That is, an ideal null sphere is a point (a, B, c) in projective $(n + 2)$ space, lying on the quadric $ac - B^2 = 0$, but not corresponding to a normal Lorentz null sphere. The totality of all normal and ideal null spheres is then in one-to-one correspondence with the indicated quadric. It is convenient on occasion to take this quadric in the alternative form

$$ac - B^2 = \xi_{-1}^2 + \xi_0^2 - \xi_1^2 - \cdots - \xi_{n+1}^2,$$

as is possible in view of the signature of the quadratic form $ac - B^2$ for suitable real linear combinations ξ_j of a, B, c; these will be chosen to lead to new coordinates that are infinitesimally synchronous with the Minkowski coordinates.

Specifically, we shall choose

$$a = \tfrac{1}{2}(\xi_{-1} + \xi_{n+1}), \qquad c = 2(-\xi_{-1} + \xi_{n+1}), \qquad B = (\xi_0, \ldots, \xi_n);$$

equivalently, the ξ_j may be expressed in terms of a, $B = (b_1, \ldots, b_n)$, and c as

$$\xi_{-1} = a - \tfrac{1}{4}c, \qquad \xi_{n+1} = a + \tfrac{1}{4}c, \qquad \xi_j = b_j \qquad \text{for} \quad 0 \leqq j \leqq n.$$

Proceeding now as earlier, we introduce the sphere coordinates u_j by the equation

$$(u_{-1}, u_0, u_1, \ldots, u_{n+1}) = (\xi_{-1}^2 + \xi_0^2)^{-1/2}(\xi_{-1}, \xi_0, \ldots, \xi_{n+1}).$$

Here we could use either sign on the indicated square root, but normalize to make the point $X = 0$ correspond to the point $(1, 0; 0, \ldots, 0, 1)$ in $S^1 \times S^n$. This gives the earlier indicated equations (where $X = (x_0, \ldots, x_n)$):

$$u_{-1} = p(1 - \tfrac{1}{4}X^2),$$

$$u_j = px_j \qquad (0 \leqq j \leqq n),$$

$$u_{n+1} = p(1 + \tfrac{1}{4}X^2),$$

where

$$p = [(1 - \tfrac{1}{4}X^2)^2 + x_0^2]^{-1/2}.$$

We could now introduce angular coordinates on the space $S^1 \times S^n$, but will define only the most important two, τ and ρ, by the equations $u_{-1} = \cos\tau$, $u_0 = \sin\tau$; $\tau = \tan^{-1}(u_0/u_{-1})$, which leave τ undefined modulo 2π; and

$$\cos\rho = u_{n+1}, \qquad x_1^2 + \cdots + x_n^2 = \frac{4\sin^2\rho}{(\cos\tau + \cos\rho)^2}, \qquad 0 \leqq \rho \leqq \pi.$$

In the particular case $n = 3$ which is physically crucial, it is possible to give a useful representation of the present space–time splitting in terms of the unitary group. Recall the representation

$$H = \begin{pmatrix} t + x & y + iz \\ y - iz & -x + t \end{pmatrix}$$

for the Hermitian matrix X corresponding to the given point (t, x, y, z) of M, and define

$$U = (1 + \tfrac{1}{2}iH)/(1 - \tfrac{1}{2}iH).$$

Then U is in the unitary group $U(2)$, and the point 0 in M corresponds to the unit matrix I in U. More specifically, U has the form

$$U = (1 - it - \tfrac{1}{4}d)^{-1}\begin{pmatrix} 1 + ix + \tfrac{1}{4}d & i(y + iz) \\ i(y - iz) & 1 - ix + \tfrac{1}{4}d \end{pmatrix},$$

where $d = t^2 - x^2 - y^2 - z^2$. In terms of U, τ may be recovered in the natural way as

$$\tau = \frac{1}{2i} \log \det U = \frac{1}{2} \arg \det U,$$

which is equivalent to the equation

$$\tau = \tan^{-1}[t/(1 - \tfrac{1}{4}d)]$$

given earlier. In addition, the u_j ($j = 1, 2, 3$) are essentially tangential near I to the x_j; more exactly,

$$u_j = x_j/[(1 - \tfrac{1}{4}d)^2 + t^2]^{-1/2} \qquad (j = 1, 2, 3)$$

with U in the form

$$U = (u_0 + iu_{-1})\begin{pmatrix} u_4 + iu_1 & -u_3 + iu_2 \\ u_3 - iu_2 & u_4 - iu_1 \end{pmatrix}.$$

These u_j are identical to the spherical coordinates previously indicated, for $S^1 \times S^3$. It follows that near the origin in Minkowski space,

$$u_j = x_j + \text{terms of third or higher order} \qquad (j = 0, 1, 2, 3).$$

We now turn to the consideration of how the various basic symmetries act or are represented in terms of these parametrizations—these are notably the L_{jk}, Lorentz transformations, conformal inversion, etc. To indicate the method employed, consider the question of the action on $U(2)$ of conventional temporal translation $t \to t + s$ in Minkowski space. We note first that the indicated mapping of M into $U(2)$, while it does not cover all of $U(2)$, omits just those elements of $U(2)$ corresponding to ideal null spheres, or in terms of $U(2)$ itself, precisely those unitaries U for which -1 is an eigenvalue. As far as the correspondence between M and $U(2)$ is concerned, this is a well-known fact about the Cayley transform $H \to (1 + \frac{1}{2}iH) \times (1 - \frac{1}{2}iH)^{-1}$. That the omitted points in $U(2)$ correspond in one-to-one fashion to the ideal points in $\tilde{Q}$ follows from the precisely two-to-one representation of points in $\tilde{Q}$ in terms of the u_j, in terms of which the general element in $U(2)$ may also be represented in a matching two-to-one fashion.

5. Causal symmetries and the energy

We turn now to the consideration of the symmetries acting on the two cosmos.† We aim to give explicit expressions for the relevant symmetries, in

† We use the term *unispace* (short for universal covering space) for the universal space with the foregoing physical interpretation. Thus unispace is conformally an infinite-sheeted covering of Minkowski space augmented by a light cone at infinity. Similarly, *unitime* refers to the natural time τ in this space.

terms of the various parametrizations. These are useful in computations, and in clarifying special relativistic formalism. Before going into the matter of explicit formulas, we enumerate the key qualitative results regarding symmetries.

(1) Every causal automorphism of Minkowski space M, i.e., every Lorentz transformation or scale transformation, or product of such, extends uniquely and without singularities to a corresponding transformation on conformal space $\bar{M}$. That is, the restricted conformal group does indeed act on $\bar{M}$, where "restricted conformal group" is defined as the 11-parameter group of transformations on Minkowski space consisting of products of Lorentz with scale transformations. The reversal operations (time/space/total) also continue to act, on all of $\bar{M}$, without singularities.

(2) In addition, conformal inversion, formally the transformation $Q: X \to 4X/X^2$, although singular on M, becomes an analytic everywhere-defined transformation on all of $\bar{M}$. Together with the action on $\bar{M}$ of the restricted conformal group, Q generates a 15-parameter Lie group of transformations on $\bar{M}$; this is the action of $O(n, 2)$ earlier derived.

(3) Space rotations around an observer in Minkowski space correspond to space rotations in $S^1 \times S^n$ around the corresponding point—i.e., the space rotations are essentially the same in the two models. However, the temporal generators (i.e., energies) are basically different. The unispace generator is *strictly greater than the Minkowski energy*, and differs from it essentially by εQEQ, where ε is a small constant and E is the Minkowski energy operator.

Consider now the two temporal evolution groups. The transformation $T: t \to t + s$ in Minkowski space can be represented as a transformation either on $S^1 \times S^n$, or in the case $n = 3$, as a transformation on $U(2)$. Taking the latter representation first, with the correspondence

$$H = \begin{pmatrix} t + x & y + iz \\ y - iz & t - x \end{pmatrix},$$

T carries $H \to H + sI$. The corresponding element U in $U(2)$ to H is

$$U = \frac{1 + \frac{1}{2}iH}{1 - \frac{1}{2}iH}; \qquad \text{whence} \quad H = -2i\frac{U - I}{U + I}.$$

Representing T as a transformation on $U(2)$, it follows by a simple computation that its action is

$$U \to \frac{(1 + \frac{1}{4}is)U + \frac{1}{4}is}{-\frac{1}{4}isU + (1 - \frac{1}{4}is)}.$$

This is a transformation in the standard action of the group $SU(2, 2)$ on $U(2)$. Any element of $SU(2, 2)$ can be represented in natural fashion in the form $\begin{pmatrix} A & B \\ C & D \end{pmatrix}$ where A, B, C, D are suitable 2×2 matrices; in these terms the standard action is

$$U \to (AU + B)(CU + D)^{-1}.$$

Denoting the Lie algebra of this group as $su(2, 2)$, this can be identified with the 4×4 skew Hermitian matrices relative to the Hermitian form $z_1 \bar{z}'_1 + z_2 \bar{z}'_2 - z_3 \bar{z}'_3 - z_4 \bar{z}'_4$. With this identification, the generator of Minkowski temporal translation is then represented by the matrix (where I denotes the 2×2 identity matrix)

$$\frac{d}{dt} \to \frac{1}{4}\begin{pmatrix} I & I \\ -I & -I \end{pmatrix}.$$

On the other hand, the unitime group is the group $U \to e^{is}U$, having the generator

$$\frac{d}{dt'} \to \frac{1}{2}\begin{pmatrix} I & 0 \\ 0 & -I \end{pmatrix}.$$

These one-parameter groups are not only distinct, but nonconjugate, within the group $SU(2, 2)$, as shown by the difference between the spectra of the two generators, which are respectively continuous and discrete, in relevant representations (cf. Segal, 1967a). This is evident also from the fact that one of the matrices is singular and the other nonsingular. However, *their local actions at a fixed point of space are very close, differing only by terms of order s^3 and higher, where s is the elapsed time.* To show this, it is by homogeneity no essential loss of generality to choose the fixed point of space as the origin in Minkowski space, or as the unit I in $U(2)$. (The correspondence between M and $U(2)$ depends on the point of reference; however, that may be arbitrarily designated as the origin in Minkowski space, in view of the Lorentz invariance of M, $\bar{M}$, and their relation.) The unitime group sends $I \to e^{is}I$; the special relativistic temporal group sends

$$I \to \frac{1 + \frac{1}{2}is}{1 - \frac{1}{2}is} I;$$

these differ by $O(s^3)$.

It is thus evident that the special relativistic energy appears relatively complicated in the $U(2)$ formalism; this is also the case in the $S^1 \times S^3$ formalism. On the other hand, the unienergy appears relatively complicated in the Minkowski space formalism. To compute the action of unitemporal evolution in the Minkowski picture, we must transfer the action $U \to e^{is}U$ to an action on the corresponding H, and hence to an action on (t, x, y, z). Now, unitemporal evolution sends

$$U \to e^{is}U.$$

Since

$$H = -2i\frac{U - I}{U + I}, \qquad H \to -2i\frac{e^{is}U - 1}{e^{is}U + 1};$$

now, expressing U in terms of H again, it results that

$$H \to H' = \frac{H + 2\tan(s/2)}{1 - \frac{1}{2}H\tan(s/2)}.$$

We now relate the Minkowski energy to conformal inversion and the unienergy, and compute the latter in this way. We define conformal inversion, to be designated Q, as the transformation

$$Q: X \to 4X/X^2;$$

this is not everywhere defined on M, but extends naturally to an everywhere-defined transformation on the larger space $\bar{M}$, as follows. In terms of the coordinates (a, B, c), Q is the map $(a, B, c) \to (c, 4B, 16a)$, as follows on substitution in the equation for a Lorentz sphere. It results that on the ξ_j, Q acts as

$$Q: \xi'_{-1} = -\xi_{-1}; \qquad \xi'_j = \xi'_j \qquad (j \geqq 0),$$

and on the u_j,

$$u_{-1} \to -u_{-1}; \qquad u_j \to u_j \qquad (j \geqq 0).$$

Thus, in unispace, conformal inversion affects only the time, not the space component. It is only the identification of antipodal points in $S^1 \times S^n$ which gives the transformation its spatial character on $\bar{M}$ or $U(2)$. For example, the transformation

$$(1, 0; 0, \ldots, 1) \to (-1, 0; 0, \ldots, 1)$$

effected by conformal inversion on the origin (observational location) is purely temporal; but $(-1, 0; 0, \ldots, 1)$ represents the same point of conformal space as does its multiple by -1, i.e., $(1, 0; 0, \ldots, -1)$, which can be considered to be the image under purely spatial inversion of the origin.

We should distinguish between proper unispace and the locally identical but acausal space $S^1 \times S^n$, which becomes conformal space upon identifying antipodal points. In contrast, the former space is $R^1 \times S^n$ and is mapped upon $S^1 \times S^n$ by the transformation $(t, u) \to (e^{it}, u)$, and thence upon conformal space, in which Minkowski space is properly contained. It is important to note that Q extends naturally not only from an improper transformation in M to a proper one in $\bar{M}$, but also corresponds to proper transformations in both $S^1 \times S^n$ and $R^1 \times S^n$. In the case of $S^1 \times S^n$, the transformation is $Q^{(2)}$: $(\lambda, u) \to (-\lambda^{-1}, u)$ where λ is a complex number of absolute value 1 representing the S^1 component. The temporal character of conformal inversion as a transformation on $R^1 \times S^n$ is particularly clear: it is $Q^{(\infty)}$: $(t, u) \to (\pi - t, u)$. Here we have adopted the notational device of adding a superscript to designate the space, whether conformal space, its twofold covering space $S^1 \times S^n$, or its ∞-fold covering $R^1 \times S^n$. (There are similar coverings for each subgroup of the center Z of the group $\widetilde{SO}(2, 4)/Z_2$ of all causality-preserving transformations on $\tilde{M}$; Z is an infinite cyclic group which is generated by the transformation $(t, u) \to (t + \pi, -u)$. These additional coverings may be relevant to elementary particle considerations, but not directly to the astrophysical considerations of present concern. Hence they will be ignored in the following.)

It thus appears that conformal inversion differs from time reversal in unispace by the transformation on $R^1 \times S^n$, $(t, u) \to (t + \pi, u)$. The latter transformation is evidently contained in the one-parameter group $t \to t + s$, and so is continuously connected to the identity transformation. Since time reversal is represented by an antiunitary transformation in any positive energy particle model, it follows that *conformal inversion is represented by an antiunitary operator in any positive-energy particle model.*

In the case $n = 3$, it follows from the correspondence between $S^1 \times S^3$ and $U(2)$ that conformal inversion acts as follows on $U(2)$:

$$Q\colon U \to -\frac{U}{\det(U)}.$$

Let us now consider the transformation properties of the energies, Minkowski and universal, under Q. It is evident that the Minkowski temporal evolution generator $\partial/\partial t$ does not commute with Q. A straightforward computation gives

$$Q\frac{\partial}{\partial t}Q = -\frac{1}{4}(t^2 + x^2 + y^2 + z^2)\frac{\partial}{\partial t} - \frac{t}{2}\left(x\frac{\partial}{\partial x} + y\frac{\partial}{\partial y} + z\frac{\partial}{\partial z}\right).$$

On the other hand, if τ denotes time in unispace, then $\partial/\partial\tau$, the generator of unispace temporal evolution, is carried by Q into $-\partial/\partial\tau$. Indeed, $\partial/\partial\tau = u_{-1}(\partial/\partial u_0) - u_0(\partial/\partial u_{-1})$; on sending $u_{-1} \to -u_{-1}$ and $u_0 \to u_0$, this is evidently reversed. This shows explicitly that in any particle model in which the unienergy is positive, conformal inversion must be represented by an antiunitary operator.

Let us now compute how $\partial/\partial\tau$ appears in terms of Minkowski coordinates t, x, y, z. To do this, note that since $\exp(a\,\partial/\partial\tau)$ sends $U \to e^{ia}U$, for $U \in U(2)$, the corresponding action on H is as earlier computed,

$$H \to \frac{H + 2\tan(a/2)}{1 - \frac{1}{2}H\tan(a/2)} = H + aI + (\tfrac{1}{4}a)H^2 + O(s^2).$$

Thus, $\partial/\partial\tau$ carries H into $I + \frac{1}{4}H^2$. Recalling now that

$$H = \begin{pmatrix} t + x & y + iz \\ y - iz & t - x \end{pmatrix},$$

it follows that

$$\frac{\partial}{\partial\tau} = \frac{\partial}{\partial t} + \frac{1}{4}(t^2 + x^2 + y^2 + z^2)\frac{\partial}{\partial t} + \frac{t}{2}\left(x\frac{\partial}{\partial x} + y\frac{\partial}{\partial y} + z\frac{\partial}{\partial z}\right).$$

Comparing the latter equation with that for $Q(\partial/\partial t)Q$, it results that

$$\frac{\partial}{\partial\tau} = \frac{\partial}{\partial t} - Q\frac{\partial}{\partial t}Q.$$

Corollary In any particle or field model that is (a) invariant under the group $O(n, 2)$ (or any locally identical group), (b) such that the special relativistic energy is positive, the unienergy is also positive, and exceeds (in all states) the special relativistic energy.

Proof Let U be any unitary–antiunitary representation in Hilbert space of the group $O(n, 2)$, and conformal inversion Q, which in particular is represented by an antiunitary operator Q, as earlier noted. By the foregoing equation,

$$U\left(\frac{\partial}{\partial\tau}\right) = U\left(\frac{\partial}{\partial t}\right) - U\left(Q\frac{\partial}{\partial t}Q\right),$$

where we denote also by U the natural extension of U to infinitesimal operators of the group. Now

$$U\left(Q\frac{\partial}{\partial t}Q\right) = U(Q)U\left(\frac{\partial}{\partial t}\right)U(Q).$$

The assumption that the special relativistic energy is positive means that $i^{-1}U(\partial/\partial t)$ is a positive (as well as automatically Hermitian) operator. Since $U(Q)$ is antiunitary, it carries any one-parameter unitary group with positive generator into one with a negative generator; i.e., $i^{-1}U(Q)U(\partial/\partial t)U(Q)$ is a negative Hermitian operator. Thus the unienergy $-iU(\partial/\partial\tau)$ is the sum of the relativistic energy and the positive operator $iU(Q\ \partial/\partial t\ Q)$.

As a further example, let us compute some of the generators L_{ij} of the conformal group in terms of space–time coordinates x, y, z, t. The general procedure is as follows. For $k = 0, 1, 2, 3$,

$$L_{ij}x_k = L_{ij}\left(\frac{2\xi_k}{\xi_{-1} + \xi_4}\right) = 2(\xi_{-1} + \xi_4)^{-1}L_{ij}\xi_k - 2\xi_k(\xi_{-1} + \xi_4)^{-2}L_{ij}(\xi_{-1}),$$

whence in terms of the x_k,

$$L_{ij} = 2\sum_k [(\xi_{-1} + \xi_4)^{-1}L_{ij}\xi_k - (\xi_{-1} + \xi_4)^{-2}\xi_k L_{ij}(\xi_{-1} + \xi_4)]\frac{\partial}{\partial x_k}.$$

Evidently, $L_{ij}\xi_k = \varepsilon_i\,\delta_{jk}\,\xi_{-i}\varepsilon_j\,\delta_{ik}\,\xi_j$. Thus, e.g., setting $e = t^2 + x^2 + y^2 + z^2$,

$$L_{-1,0} = \left(1 + \frac{e}{4}\right)\frac{\partial}{\partial t} + \frac{t}{2}\left(x\frac{\partial}{\partial x} + y\frac{\partial}{\partial y} + z\frac{\partial}{\partial z}\right),$$

$$L_{0,4} = \left(1 - \frac{e}{4}\right)\frac{\partial}{\partial t} - \frac{t}{2}\left(x\frac{\partial}{\partial x} + y\frac{\partial}{\partial y} + z\frac{\partial}{\partial z}\right),$$

$$L_{-1,4} = -\left(t\frac{\partial}{\partial t} + x\frac{\partial}{\partial x} + y\frac{\partial}{\partial y} + z\frac{\partial}{\partial z}\right).$$

It should now begin to be visible how a theoretical analysis could be conducted by an observer, based on either the curved (unispace, $R^1 \times S^3$) or flat (Minkowski, $R^1 \times R^3$) local decompositions of space–time into space and time components. In both cases, all the fundamental physical laws on which conventional reduction and analysis of observation are based remain valid. Specifically, in either case:

(a) there is a proper global notion of causality in the Cosmos; and locally the notions are identical in the two models;

(b) conservation of energy, angular momentum, etc. are valid;

(c) Lorentz invariance holds; given any two future directions of a point, corresponding to relatively accelerated observers, they are related by a global causality-preserving transformation;

(d) the energy is positive;

(e) there are essentially unique measures of temporal duration and of spatial distance, which are invariant under the respective underlying symmetry groups (unicity within a scale factor);

(f) there is finite propagation velocity for the causal structure relative to the time and distance measures in (e).

Physically, only one of these analyses can be globally valid, however. For example, if global conservation of energy is valid in one analysis, it cannot be valid in the other, for the respective energy operators do not commute. At first glance, it might appear that empirical confirmation of special relativity precluded the empirical validity of the curved formulation of the local space–time splitting. It develops, however, that as regards direct measurements the local unispatial analysis differs negligibly from the special relativistic analysis; it is only at extreme distances that significant differences emerge. This arises basically from two circumstances: (a) the unispatial space–time splitting is not only tangent to the special relativistic space–time splitting at the observer's location, but has the remarkable feature of agreeing with it within terms of third or higher order; (b) the basic distance scale, set in the unispace theory by the radius of the universe, is such that the times and distances involved in classical local measurements are ultramicroscopic.

In order to determine the distance scale, we must evaluate in conventional units the "radius of the universe" R, i.e., the radius of the S^3 component of unispace. At the present time, this can be deduced only from redshift measurements; this is natural since no other measurements are known to involve very great distances. It will be seen that the two time scales differ only by at most ~ 1 part in 10^{19} in the course of a year (or less), and similarly for distance scales (scaled by the velocity of light); the difference is thus well beyond the limits of present experimental capabilities. It is only by indirect measurements at extreme distances that the difference between the two models of an observer is empirically perceivable.

6. The redshift

Let us then provisionally adopt the unispace cosmos, and seek to analyze free propagation over very long times, and its effect on the measurement of frequency of light. The determination of the wavelength of light is very much of a local matter in practice. Conventionally the frequency is represented by the operator $i^{-1}\,\partial/\partial t$. We now have at hand an alternative possibility $i^{-1}\,\partial/\partial\tau$, where τ is the time in unispace. In the absence of any observed phenomenon such as the redshift, it might perhaps seem equally natural to represent the frequency by this alternative; there would, however, then be no apparent means of determining the distance scale, i.e., of measuring τ in natural units. In addition, the whole procedure of local measurement in the vicinity of a fixed observer is based on flat geometry. These considerations give a certain preferential basis for the flat energy $i^{-1}\,\partial/\partial t$, independently of the results of observational extragalactic astronomy; but what is really strongly indicative, indeed conclusive, is the fact of the observed redshift, which could not exist in the unispace cosmos if the alternative representation for the observed frequency were valid. We are led thereby to postulate that anthropomorphically possible local measurements are represented theoretically by the *flat* rather than *curved* dynamical variables; while on the other hand, the "true" nonanthropomorphic dynamics and analysis is curved (in the fashion appropriate to the unispace cosmos) rather than flat. That is, we measure the flat dynamical variables; but the Universe in the large runs on the curved basis, which agrees only instantaneously with the flat one. This postulate is provisional, pending the derivation of an effective treatment of redshift laws, etc., from it. We now begin this treatment, which will be found to agree with observation.

Among the dynamical variables that are chiefly involved in measurements are the space and time coordinates: the energy, angular momentum, and other particle quantum numbers. Since there is no direct means of observation of extragalactic distances, and the effect on galactic distances and coordinates is negligible, there is no apparent present possibility of distinguishing the theory by measurements of the coordinates. The angular momentum is the same in both the flat and curved cosmos, as it develops. However, the most basic of the dynamical variables, the energy, is affected in an observable fashion.

Let H_0 denote the dynamical variable $-i(\partial/\partial t)$ which has been postulated to represent theoretically the result of a local measurement of frequency. We are particularly interested in the case of light, its frequency being measured by the usual optical methods. According to the unispace theory, this dynamical variable is not the true energy, but only appears to be so infinitesimally at the location of an observer. (However, there is a scale factor between the two energies, and if synchronous at one location the scale factor will differ from unity at other locations.) The special relativistic energy is, therefore, not conserved, just as the energy relative to one Lorentz frame is not conserved relative to the temporal development in another frame. In the latter case, the difference in energy is relatively gross, being for small times s of the order of const $\times$ s, and so should be readily observable. In the present case, however, the unispatial frame is tangent to the flat frame, and defines an identical Lorentz frame at the observer's location; it is only the nonlinearity of the relation between the two cosmos and their respective groups of temporal displacement which causes a discrepancy in the energies. As a result, the extent of nonconservation of H_0 is proportional to s^2, rather than to s, for small values of s, within terms of higher order.

In order to obtain an exact expression, it is necessary to compute explicitly the dynamical variable representing the frequency after passage of a time s, i.e., the operator $H_0(s) = e^{-isH} H_0 e^{isH}$, where H is the true (conserved) energy, given in the unispace cosmos as $-i(\partial/\partial\tau) = -iL_{-1,0}$. Thus

$$H_0(s) = e^{-sL_{-1,0}} \frac{1}{2i}(L_{-1,0} + L_{0,4})e^{isL_{-1,0}}$$

$$= \frac{1}{2i} L_{-1,0} + \frac{1}{2i} e^{-sL_{-1,0}} L_{0,4} e^{sL_{-1,0}}.$$

Now $L_{-1,0}$, $L_{0,4}$, and $L_{-1,4}$ generate a three-dimensional subgroup of $O(4, 2)$; in particular

$$[L_{-1,0}, L_{0,4}] = L_{-1,4}, \qquad [L_{-1,0}, L_{-1,4}] = -L_{0,4}.$$

It follows that

$$e^{-sL_{-1,0}}L_{0,4}e^{sL_{-1,0}} = A(s)L_{-1,4} + B(s)L_{0,4},$$

where $A(0) = 0$, $B(0) = 1$. To evaluate $A(s)$ and $B(s)$, we take first and second derivatives in the foregoing equation. It results that

$$-e^{-sL_{-1,0}}[L_{-1,0}, L_{0,4}]e^{sL_{-1,0}} = A'(s)L_{-1,4} + B'(s)L_{0,4};$$

evaluating the commutator, it follows that

$$-e^{-sL_{-1,0}}L_{-1,4}e^{sL_{-1,0}} = A'(s)L_{-1,4} + B'(s)L_{0,4},$$

whence $A'(0) = -1$, $B'(0) = 1$. Differentiating again, it follows similarly that

$$-e^{-sL_{-1,0}}L_{0,4}e^{sL_{-1,0}} = A''(s)L_{-1,4} + B''(s)L_{0,4},$$

which implies that $A''(s) + A(s) = 0 = B''(s) + B(s)$. It results that

$$A(s) = -\sin s, \qquad B(s) = \cos s,$$

whence

$$\begin{aligned} H_0(s) &= \frac{1}{2i}[L_{-1,0} - \sin sL_{-1,4} + \cos sL_{0,4}] \\ &= \frac{1+\cos s}{2i}\frac{\partial}{\partial t} + \frac{1-\cos s}{2i}Q\frac{\partial}{\partial t}Q \\ &\quad + \frac{\sin s}{2i}K, \qquad (K = -L_{-1,4}). \end{aligned}$$

Now consider how this change in the operator $i^{-1}\,\partial/\partial t$ representing frequency measurement, over the time interval of duration s, is reflected in a measurement of frequency of a light wave, initially of a fixed frequency ν. The wave function Ψ has then the feature that

$$\frac{1}{i}\frac{\partial\Psi}{\partial t} = \nu\Psi, \qquad \text{i.e.,} \qquad H_0(0)\Psi = \nu\Psi,$$

at the point of emission, which may be taken as the origin. At the later point P at which the frequency is measured, say of coordinates (s, u), where s is the unitime and u the position in space S^3, the observed frequency ν' will be given by the equation

$$H_0(s)\Psi = \nu'\Psi.$$

However, while ν will be an exact value for the frequency $H_0(0)$, i.e., Ψ is a stationary state for H_0, there is no a priori reason for ν' to be an exact value for $H_0(s)$; Ψ need not be an exact stationary state for $H_0(s)$. Thus the frequency will be shifted from ν to ν', while at the same time a certain dispersion, effecting a corresponding line broadening, may be introduced into the frequency measurement. In order to compute ν' and its dispersion, we must know explicitly the wave function Ψ. The simplest reasonable postulate is that it is a plane wave of frequency ν. Neglecting polarization, which is presently irrelevant, it has then the form

$$\psi(t, x) = e^{i\nu(t - \mathbf{x} \cdot \mathbf{k})},$$

k being fixed. Strictly speaking, this representation for the wave function is incomplete, in that it is given as a function of the special relativistic coordinates, which are not globally applicable throughout unispace. It will be seen later that it nevertheless extends in a natural and unique way to a wave function throughout the region accessible from the point of emission by a light ray. In the meantime, we shall ignore apparent questions as to behavior of the wave function at extremely remote points of the Cosmos, e.g., our antipode.

We wish first to determine $H_0(\tau)\psi$, at the point of observation, which we take to have the form (τ, u), where u is the spatial position in S^3. Since Maxwell's equations and the wave equation are conformally invariant, the properties of solutions are basically independent of whether they are analyzed from a flat or curved standpoint. In particular, light continues to be propagated along light rays of the conventional type, which however appear in unispace to have the form $(\sigma, u(\sigma))$, $0 \leqq \sigma \leqq \sigma_0$, where $u(\sigma)$ describes a great circle on S^3 with constant velocity, normalized to be 1. In particular, at the point of observation, $\tau > 0$ and the distance of u from the point in S^3 of emission, taken here as $(0, 0, 0, 1)$, is precisely τ. It is essentially the same to say that, in terms of flat coordinates, $t^2 = x^2 + y^2 + z^2$, except that this parametrization is valid only on part of unispace.

Consider first $K\Psi$ evaluated at P. Evidently

$$K\Psi = \left(x\frac{\partial}{\partial x} + y\frac{\partial}{\partial y} + z\frac{\partial}{\partial z} + t\frac{\partial}{\partial t}\right)\Psi = i\nu(t - \mathbf{x} \cdot \mathbf{k})\Psi.$$

However, $t = \mathbf{x} \cdot \mathbf{k}$ along the ray of propagation, in particular at the point P. Thus $K\Psi|_P = 0$. Next, consider $Q(\partial/\partial t)Q$ evaluated at P. As earlier determined, $Q(\partial/\partial t)Q$ is a linear combination with constant coefficients of the operators $(t^2 - x^2 - y^2 - z^2)\,\partial/\partial t$ and tK. Evidently, the first of these operators on application to Ψ yields zero along any light ray. The second vanishes at P on application to Ψ by the preceding paragraph.

Thus

$$H_0(\tau)\Psi|_P = \frac{1 + \cos\tau}{2i}\frac{\partial\Psi}{\partial t}\bigg|_P = \frac{1 + \cos\tau}{2}\nu\Psi\bigg|_P \qquad \text{i.e., at } P,$$

$$H_0(\tau)\Psi = \nu'\Psi \qquad \text{with} \quad \nu' = \frac{1 + \cos\tau}{2}\nu.$$

This means there is an expected frequency-independent redshift in the amount

$$z = \frac{1 - \cos \tau}{1 + \cos \tau} = \tan^2\left(\frac{\tau}{2}\right).$$

However, for such a redshift of a discrete frequency level to be observable, it is necessary that the dispersion in the expected frequency be relatively small (of the order of no more than some angstroms). A bound on the dispersion indicating that this is the case can be obtained without detailed computation in the following manner.

By general quantum phenomenological principles, the variance σ^2 of a dynamical variable X in a given state may be expressed as follows. Let E denote the expectation value functional for the state; i.e., for a given dynamical variable Y, $E(Y)$ is the expectation of Y in the state. Then

$$\sigma^2 = E(X^2) - E(X)^2.$$

Let us apply this to the state ψ and the dynamical variable $H' = H(\alpha)$. We have seen that near the point P of observation

$$H'\psi \sim \nu'\psi.$$

The variance σ^2 may then be computed from the equation

$$H'^2\psi \sim (\nu'^2 + \sigma^2)\psi$$

near P. Now $H'\psi = a\psi$, where a is a certain function on space–time; it follows that

$$H'^2\psi = aH'\psi + (H'a)\psi = a^2\psi + (H'a)\psi,$$

inasmuch as H' is a homogeneous first-order linear differential operator. It follows that

$$\sigma^2 = H'a = \nu[H_0(\alpha)(t - x \cdot k)]_p.$$

This shows that σ is of the order of $\nu^{1/2}$, and hence negligible relative to ν for large ν.

The frequencies involved here (e.g., for an observed wavelength of 21 cm or less) are indeed quite large, especially in the units here in question. In these units, π units of time are required for light to traverse the distance from any point of space S^3 to its antipode. Assuming this distance to $\gtrsim 10^9$ ly (cf. Chapter IV) and the frequency ν to correspond to an emitted wavelength ≤ 20 cm, this gives $\nu^{1/2}/\nu \lesssim 10^{-12}$, an entirely negligible (unobservable) dispersion. This applies in fact to an arbitrary stationary state. Within the plane wave approximation, the dispersion vanishes, i.e., ψ is effectively a stationary state of H' near the point of observation. But it is not at all an eigenvector for H' throughout the Cosmos.

In fundamental principle, expectation values and dispersions are defined by integration over the entire Cosmos; in practice, one analyzes only the behavior of the wave function in the vicinity of the points of interest. To within the approximation represented by the use of plane waves rather than normalizable photon wave functions, this accurately reflects the observational situation. In effect, one cuts off the defining integrals at a distance, say of the order of 1 ly, which is far beyond the limits of direct observation, but sufficiently small that $H_0(\alpha)^n\psi \sim v'^n\psi$ $(n = 1, 2)$ out to this distance, within observational accuracy. Without such a cutoff, the defining integrals over the entire Cosmos would, in the case of plane wave, be divergent.

This cut-off may be made rigorous and the entire redshift computation carried out within the Hilbert space of normalizable photon wave functions at the cost of some analytical complication. Since the Hilbert space analysis is the basis of the correlation of the Heisenberg picture just adopted with the Schrödinger picture, it seems useful to develop it. To do so, the photon Hilbert space must be set up explicitly. Within the scalar approximation already adopted, a photon may be represented by a solution φ of the wave equation, of the form

$$\varphi(X) = \int_{K^2=0} e^{iX\cdot K} f(K)\, d\mu(K),$$

where $X = (x_0, x_1, \ldots, x_n)$, $K = (k_0, k_1, \ldots, k_n)$, $X \cdot K = x_0 k_0 - x_1 k_1 - \cdots - x_n k_n$, n is the number of space dimensions, and $d\mu(K) = dk_1 \cdots dk_n/|k_0|$. The inner product between two such wave functions is given by the equation

$$\langle \varphi_1, \varphi_2 \rangle = \int_{K^2=0} f_1(K) \bar{f}_2(K)\, d\mu(K).$$

This inner product is invariant under all orthochronous conformal transformations and is uniquely determined, within a constant factor, by this property.

According to the Heisenberg form of quantum mechanics, the operator $H_0(s) = e^{-isH} H_0 e^{isH}$ representing the relativistic energy at time s has expectation value $\langle H_0(s)\varphi, \varphi \rangle$ and variance $\langle H_0(s)^2\varphi, \varphi \rangle - \langle H_0(s)\varphi, \varphi \rangle^2$ if the photon is in the state φ, normalized by the condition that $\langle \varphi, \varphi \rangle = 1$.

In the Schrödinger picture in which the state changes but the dynamical variables remain unchanged, thc photon state after time s is $\varphi_s = e^{isH}\varphi$, and the expectation value and variance of the relativistic energy H_0 in this state are given by the expressions $\langle H_0 \varphi_s, \varphi_s \rangle$ and $\langle {H_0}^2 \varphi_s, \varphi_s \rangle - \langle H_0 \varphi_s, \varphi_s \rangle^2$, which are equal to those earlier given by virtue of the unitarity of the operators e^{isH}. Thus, as is well known, the two pictures give physically equivalent results. The computation of the redshift and its dispersion evidently depends on the evaluation of the inner products $\langle H_0\varphi, \varphi\rangle$, $\langle H_1\varphi, \varphi\rangle$, $\langle K\varphi, \varphi\rangle$, $\langle H_0{}^2\varphi, \varphi\rangle$, $\langle H_0\varphi, H_1\varphi\rangle$, etc.

Real photons may equivalently be represented by positive frequency wave functions, i.e., complex-valued wave functions φ that satisfy the wave equation and have vanishing negative frequency components, instead of wave functions φ that are real in physical space. Since the Fourier transform $F(K)$ of the latter type of wave function is hermitian, $F(-K)=\overline{F}(K)$, it is determined by its positive frequency component, on which the orthochronous conformal group acts in the same way. A positive frequency wave function can not be localized in physical space, since it consists of boundary values of an analytic function; consequently the most direct form of representation of a recently emitted (and therefore localized) photon is in terms of a real wave function. The simplest such function that seems physically relevant is a cutoff plane wave in two space–time dimensions, the one-dimensional space being defined by the direction of motion of the photon. This is of the form

$$\varphi(x_0, x_1)=f(\nu(x_1-x_0)), \qquad f(x)=\begin{cases}1+\cos x & |x|\le p\\ 0 & |x|\ge p\end{cases}$$

where p is of the form $p=(2r+1)\pi$, r being an integer; this is simply a plane wave of frequency ν, which is cut off smoothly beyond $2r+1$ oscillations.

Without explicit computation, it follows that the ν-dependence of the relevant integrals is as follows, after normalization of the wave function by division by $\langle\varphi, \varphi\rangle^{1/2}$:

$$\begin{array}{lll}\langle H_0\varphi, \varphi\rangle \propto \nu & \langle H_1\varphi, \varphi\rangle \propto \nu^{-1} & \langle K\varphi, \varphi\rangle \propto \nu^0\\ \langle H_0{}^2\varphi, \varphi\rangle \propto \nu^2 & \langle H_1{}^2\varphi, \varphi\rangle \propto \nu^{-2} & \langle K^2\varphi, \varphi\rangle \propto \nu^0\end{array}$$

By virtue of the scale on which ν is measured, according to which a typical value is $\gtrsim 10^{26}$ (the value for the 21 cm line, assuming the radius R of the universe is $\gtrsim 100$ Mpc, the figure resulting from conventional estimates of the distance and redshift of Virgo galaxies), together with the Schwarz inequalities: $|\langle H_0\varphi, H_1\varphi\rangle| \le \|H_0\varphi\|\,\|H_1\varphi\|$, etc., the only possible nontrivial contribution to a dispersion in redshift (i.e., a contribution of order ν) can come from the terms involving H_0.

Explicit computation of these terms gives the results: $\langle H_0\varphi, \varphi\rangle = \nu(1+O(\log p/p)$; $\langle H_0{}^2\varphi, \varphi\rangle = \nu^2(1+O(\log p/p)$. Thus the redshift dispersion is $\nu O(\log p/p)$ and vanishes for an infinite plane wave. For a cutoff plane wave which is $\gtrsim 1$ light-second in extent, and ≤ 21 cm in wavelength, the dispersion is $\lesssim 1$ part in 10^{17}, and hence quite remote from observation. It follows also that the superrelativistic component of recently emitted radiation within the galaxy is negligable. But old, delocalized radiation, no longer approximately an eigenstate of H_0—although necessarily resoluble into such—can be highly energetic, particularly the very low frequency components, since $\langle H_1\varphi, \varphi\rangle$ varies inversely with ν. Explicit computation shows that the proportionality factor has the form $\kappa p^2(1 + O(\log p/p))$, where κ is an absolute constant of order 1; i.e., the superrelativistic energy varies approximately directly with the square of the diameter of the region of support. This result is naturally to be compared with the approximately quadratic rate at which the superrelativistic component of the energy of a freely propagated photon builds up according to the redshift law earlier derived.

7. Local Lorentz frames

Having thus derived an apparent redshift, let us now elucidate the physical connection between the flat and curved dynamical observables. This is necessary in order to predict results of other types of measurement. A central hypothesis used in the derivation is that an anthropomorphically measurable local observable is represented theoretically by a *flat* dynamical variable at the point of observation P_0. These flat dynamical variables are mathematically definable in a large region of the Cosmos, far from the point P_0, but at other points they are mathematically distinct from the corresponding flat dynamical variables, expressible in constant coefficients in terms of the local anthropomorphic Lorentz frame. At any point P, the latter frame is the unique Lorentz frame—unicity only within a scale factor, an important point to be further discussed—which is tangential to the globally given curved unispatial frame. This is the unique such unispatial frame which at the point of observation P_0 is tangential to the Lorentz frame of measurement at P_0.

A stationary anthropomorphic observer at a point P' should then see events as taking place in this tangential Lorentz frame. These frames vary in a well-defined way with the point P, and from a conventional Minkowskian point of view are in relative motion. The latter motion is entirely virtual; the Cosmos is stationary from the curved observational standpoint, which is, however, anthropomorphically indirect (i.e., accessible via redshift–magnitude observations, etc., and their theoretical interpretation). In other terms, the true driving physics is cosmologically stationary, but the Cosmos may appear in motion due in part to the theoretical analysis employed and in part to inherent restrictions on the mode of observation enforced by anthropomorphic and/or microscopic limitations.

It is instructive to compute explicitly the relation between the anthropomorphic Lorentz frames at two different points of the Cosmos. Setting $\hbar = c = 1$ leaves open the distance scale in Minkowski space. In unispace, we employ the natural distance scale, that in which the radius R of space S^3 is unity. It will be convenient to define the anthropomorphic distance scale by a constant R which expresses the ratio between a local distance as measured in Minkowski space and as measured in unispace; this scale $R(P)$ may, for the present, vary with the point P of the Cosmos. Taking the point P_0 of observation as the origin in Minkowski space, as is no essential loss of generality, and setting R_0 for the local distance scale, the Minkowski coordinates x_j are related to the unispatial coordinates u_k as follows:

$$x_j = 2u_j R_0 (u_{-1} + u_4)^{-1},$$

in the vicinity of the point P_0. In the vicinity of a different point P_1, the local Minkowski coordinates x'_j which are tangential to the unispatial coordinates at P'—i.e., the x'_j vanish at P_1, and $dx'_j = R_1\, du_j$ ($j = 0, 1, 2, 3$) at P_1—are nonlinearly related to the Minkowski coordinates x_j. Of particular interest is the case in which P_0 and P_1 are relatively lightlike, e.g., P_0 is the point of emission of light and P_1 is the point at which it is observed. By making a suitable Lorentz transformation, it can be assumed that the x_3 and x_4 coordinates of P_1 vanish, so that one is in an essentially two-dimensional spatio-temporal situation. This simplifies the discussion, and serves to illustrate the useful simple form of the general theory in which space–time is two dimensional.

In the two-dimensional case, unispace may be fully parametrized by the angles τ and ρ, defined by the equations

$$u_{-1} = \cos\tau, \qquad u_0 = \sin\tau, \qquad u_1 = \sin\rho, \qquad u_2 = \cos\rho.$$

The tangential Minkowski coordinates at the origin are (t, x) where

$$t = \frac{2R_0 \sin\tau}{\cos\tau + \cos\rho}, \qquad x = \frac{2R_0 \sin\rho}{\cos\tau + \cos\rho}.$$

A point P_1 that is lightlike relative to the origin has unispace coordinates of the form $\tau = \rho = \alpha$. Near this point, tangential Minkowski coordinates (t', x') are given by the equations

$$t' = \frac{2R_1 \sin(\tau - \alpha)}{\cos(\tau - \alpha) + \cos(\rho - \alpha)}, \qquad x' = \frac{2R_1 \sin(\tau - \alpha)}{\cos(\tau - \alpha) + \cos(\rho - \alpha)}.$$

From these equations it is evident that (t', x') are well-defined functions of (t, x); it will suffice here to give the Jacobian matrix $(\partial x'_j / \partial x_k)$, evaluated at the point $(0, 0)$. A simple computation shows that this has the form

$$\left(\frac{R_1}{R_0}\right) \frac{\sec^2\alpha}{2} \begin{pmatrix} 1 + \cos^2\alpha & \sin^2\alpha \\ \sin^2\alpha & 1 + \cos^2\alpha \end{pmatrix}.$$

This is the product of a scale transformation, via the factor $(R_1/R_0)\sec\alpha$ with the Lorentz transformation

$$\frac{1}{2}\sec\alpha\begin{pmatrix} 1+\cos^2\alpha & \sin^2\alpha \\ \sin^2\alpha & 1+\cos^2\alpha \end{pmatrix}.$$

This can be interpreted as a *virtual* motion of velocity dependent upon α, accompanied by a *virtual* expansion with factor $(R_1/R_0)\sec\alpha$.

A natural means of determining the distance scale R is, in the present theory, based on the assumption that the redshift is entirely due to the indicated chronometric effect, apart from possible small deviations due to intrinsic velocities, local gravitational effects, etc.; the main assumption here is that the fundamental properties of matter are the same in all parts of the universe. This is a provisional assumption; conceivably these properties vary with time, and even in a stationary universe, variations in the ages of emitting objects could introduce thereby an effect on the observed redshift. It will be found, however, that there appears to be no observational evidence for a significant effect of this nature, in the sense that all of the observations discussed in Chapter IV are consistent with the simple hypothesis of a chronometric redshift. The distance scale is constant throughout the Cosmos, on the present assumption.

The relation between the canonical Lorentz frames at different points is then unique and indicates the usual rate of time dilation by the factor $1+z$. For if the local relativistic coordinates x_j near the point O of observation are normalized so that they vanish at O, the equation

$$t=\tan^{-1}(x_0(1-x_0^2/4))$$

gives the relation between the unitime t and the observational time x_0 at O. The unitimes near any other point at rest relative to the point of observation differs only in zero point from t; this is true in particular of the unitime t' at the point E of emission; it follows that $x_0 = 2\tan[(t'-t_0)/2]$, where t_0 is the zero-point difference. Now the unitime t' was synchronous with the observational time x_0' at E at the time of emission; noting that

$$dx_0=(1+x_0^2/4)dt'; \qquad x_0^2/4=\tan^2(t/2),$$

it follows that $dx_0=(1+z)dx_0'$ is the relation between emitted and observed rates. The result just derived may plausibly be applied also to the dilation of the interval between wave crests; this provides a heuristic classical derivation of the chronometric redshift–distance law which is due to H. P. Jakobsen.

Actually, the wave function of the emitted light is not known with sufficient accuracy to warrant definitive conclusions at this time regarding time dilation. Strictly speaking, it can not be an exact stationary state of either relativistic or unitemporal time displacement, since such states can not vanish outside of bounded spatial regions. Locally, it is within observational limits, a stationary state of both energy operators. The earlier derivation of the redshift–distance law for localized states of the form $\varphi(x, t) = f(t - x)$ applies equally well to states of the form $f(\tau - \rho)$. It is essential for the argument that the function f vanish outside of a small local region; if f is a complex exponential, there is no redshift, since the state is stationary for the total energy H, and the wave function is in fact normalizable, unlike the usual relativistic plane wave. Moreover the temporal characteristics of the emission process are relevant in the determination of higher-order time dilation effects, which are not necessarily small in view of the apparent extreme nonlinearity of some of the fluctuation processes strong enough to be observed at great distances. Thus, the $1 + z$ time dilation factor should be regarded as a rough overall indication, and each particular type of process should be examined on its own merits. Rust (1974) has given some theoretical and observational evidence, suggesting that the factor may be different for supernovae time lapses. A totally different case is that of short-period quasar variability; its qualitative increase with z does not as yet appear to differ markedly from a $1 + z$ law.

8. Cosmic background radiation

From the equation $z = \tan^2(\rho/2)$, it is evident that $z \to \infty$ as $\rho \to \pi$, i.e., formally the redshift approaches totality as the propagation interval approaches a half-circuit of space. However, within the Minkowski framework, the antipode is infinitely distant, so that this total redshifting requires an infinite time relative to the local flat clock at the point of emission. In addition, there are physical circumstances which quite significantly modify the formal indication for an infinite redshift.

First, the photon wave function near the antipode of the point of emission will be almost entirely delocalized, as well as highly redshifted. It will then interact appreciably with the effective plasma formed of all galaxies and possible intergalactic matter throughout all of space. The photon will no longer be freely propagated, but may be scattered or absorbed.

Second, quantum dispersion, of the order of $\nu^{-1/2}$, becomes significant when ν becomes very small, and the description of the freely propagated wave function as "redshifted" becomes oversimplified. The frequency of the photon is no longer approximately sharp. From a flat local point of view, it is low in (relativistic) energy, and relatively high in superrelativistic energy, the total unienergy being conserved. Similarly, the linear momentum is not conserved, although an analogous unimomentum is conserved. This lack of conservation of momentum applies to the direction, as well as to the magnitude, of the momentum vector. While the *expected* momentum vector has the same as the original direction of propagation, there is a stochastic component to the direction representing the quantum dispersion, with nonvanishing contributions orthogonal to the line of sight. The effect of free propagation on the linear momentum is in fact computable by the same analysis as in the case of the relativistic energy.

For all of these reasons, the analysis of highly redshifted radiation as if it were an effectively localized plane wave packet is quite inappropriate. For sufficiently high redshifts, it should not be at all observable as radiation from a discrete source, even if originating in one, but only as background radiation. Its propagation from this point onward is probably not accurately represented by the free unitemporal evolution of a solution of Maxwell's equations in Minkowski space.

While the local dynamics of all contributions to this background radiation must be extremely complex, the general considerations of equilibrium statistical mechanics lead to a conveniently simple conclusion. All radiation may be divided into two classes: (a) the "pristine," which has made less than a half-circuit of the universe since emission; (b) "residual," the remainder. The origin of this radiation is not important for general considerations, but in the chronometric theory there is no special reason not to postulate that it arises primarily from discrete objects, and for concreteness, this may be assumed. In view of the apparent transparency of intergalactic space, the residual radiation should typically make many circuits of space before ultimately interacting with matter. The infinite time available for this low-frequency, high-dispersion radiation to accumulate implies quantitatively that it may be highly energetic, but in any event qualitatively that it is distributed in accordance with Planck's law, i.e., having a blackbody spectrum. For this law follows directly from the conservation of energy and maximization of entropy. The conservation of the unienergy is the starting point in the chronometric redshift analysis of the propagation of free photons, and its extension to all dynamical processes in the universe is tantamount to temporal homogeneity and causality.† It is a very natural and almost inevitable postulate. The maximization of entropy is implied by ergo-

† It should perhaps be emphasized that the so-called steady state theory is not at all temporally homogeneous from the present standpoint since energy in this theory is essentially ad hoc and not intrinsically definable in terms of the geometrical structure of the Cosmos. It is only for a theory of this latter type, in the presence of suitable noncyclical diffusion of energy yielding the requisite ergodicity, that the Planck law follows.

dicity, or nondeterministic mixing, which should be amply fulfilled by virtue of the stochastic character of the emissions from and motions of galaxies.

Thus the residual radiation should appear in the form of an energetic background blackbody radiation. The largely unknown absorptive characteristics of the various aggregations of matter in space, as well as of the extent of this matter itself, preclude a direct estimate of the energy density of this radiation. However, an approximate upper bound on its temperature may be estimated in terms of the energy density of starlight, and certain galaxy parameters, by neglecting all absorption except that by bright galaxies. Unless there exist large amounts of matter in presently unknown form, this upper bound may reasonably be expected to give the correct order of magnitude of the temperature of the radiation.

For such order-of-magnitude estimates, it suffices to approximate the galaxies by completely absorbing spheres of a fixed radius r. The extinction in a short time τ of propagation is the quotient of the **total** solid angle Ω subtended by all the galaxies in the spherical region of radius τ subtended from the center, by 4π. Again, Ω is sufficiently accurately estimated by placing all the galaxies at the expected distance, on the basis of spatial homogeneity, of $(\frac{3}{4})\tau$ from the point of emission, and neglecting overlapping solid angles. If μ denotes the number density of bright galaxies, the resulting extinction is consequently

$$[4\pi(3\tau/4)^2]^{-1}[\mu(4/3)\pi\tau^3]\pi r^2 = (16\pi/27)\mu r^2\tau,$$

implying an extinction of $\exp[-(16\pi/27)n\mu r^2]$ in the course of n half-circuits of space.

The flat (special relativistic) component of the pristine radiation is the space average of $(1 + z)^{-1}$ times the total pristine radiation, say S. Assuming spatial homogeneity, the distribution of z is $(2/\pi)z(1 + z)^{-2}\, dz$, which when averaged over space gives a factor of $\frac{1}{2}$. On the other hand, summing over all possible numbers of circuits, the total residual radiation amounts to

$$P\sum_{n=1}^{\infty} \exp[-(16\pi/27)\mu r^2] \sim [(16\pi/27)\mu r^2]^{-1}P.$$

Thus the ratio of the energy of the residual to that of the special relativistic pristine radiation is $\sim 0.4\mu^{-1}r^{-2}$.

In making a comparison with observation, it would be natural to identify the observed microwave background radiation with the theoretical residual radiation, and, to an adequate approximation, the starlight background with the special relativistic pristine radiation. Although quantitative astronomical discussion is being left for Chapter IV, it may here be remarked that the results concerning the background radiation are in satisfactory agreement with observation.

One of the most striking features of the observed cosmic background radiation is its apparent strong isotropy. While unexpected from a Friedmann model standpoint, there is no reason for any local anisotropy within the universal cosmos framework, apart from peculiar motions. These are probably quite small for galaxies, indeed appear of an order $\lesssim 60$ km/sec (see Chapter IV). Even on a classical basis for analysis of radiation, such slight motions should not produce presently observable anisotropy. Taking into account the quantum dispersion additionally involved in a more exact treatment could only increase the threshold of peculiar motions which would produce observable anisotropy, quite possibly to a level well above that of the Sun. On the chronometric hypothesis, anisotropy in the background radiation appears unlikely to be observed until considerably greater precision of measurement is obtainable, if ever.

9. Special relativity as a limiting case of unispatial theory

Special relativity can be regarded as a limiting case of unispatial theory, as the radius R of the universe becomes infinity, in a sense indicated by Einstein, Minkowski, and Weyl. The radius R is actually a physical constant, and the mathematical content of the formation of the limit as $R \to \infty$ requires some clarification. It can, however, be defined by analogy with the familiar cases $c \to \infty$, which leads to Galilean relativity from special relativity, and $h \to 0$ which leads to classical from quantum mechanics. We shall be relatively explicit and show how the fundamental dynamical variables of the unispace theory converge to those of special relativity.

Consider first the space–time coordinates. The physically observed coordinates are not the x_j $(0 \leqq j \leqq 3)$ of the first part of this chapter, but rather the $x'_j = Rx_j$, R being the radius of the universe. These x'_j are to be compared with the Ru_j. We have

$$Ru_j - x'_j = O(1/R) \qquad \text{as} \quad R \to \infty.$$

For

$$\begin{aligned} Ru_j - x'_j &= Rx_j[(1 - \tfrac{1}{4}d)^2 + t^2]^{-1/2} - x'_j \\ &= x'_j\left\{\left[\left(1 - \frac{d'}{4R^2}\right)^2 + \frac{t'^2}{R^2}\right]^{-1/2} - 1\right\} \\ &= x'_j O(R^{-2}). \end{aligned}$$

An even sharper and uniform estimate holds when $t'^2 + x'^2 + y'^2 + z'^2 = e'$ is sufficiently small, as would be the case out to classical macroscopic distances, according to the estimates of R in Chapter IV. Thus the suitably scaled Minkowski and universal space–time coordinates agree to within a close approximation, for moderate e'.

Consider next the dynamical variables corresponding to generators of space–time symmetries. These are specifically the energy–momentum vector, the angular momenta, the boosts, and the infinitesimal scale transformation K. There are 11 generators here in all; seven of them are the same (i.e., have the same expression in terms of the x_j' and $\partial/\partial x_j'$) in both theories; the energy–momentum vector is the quartet which are distinct between the two theories. There are, in addition, four linearly independent infinitesimal symmetries of unispace, whose action in Minkowski space as $R \to \infty$ remains to be explored.

We have the

Theorem The suitably scaled 15 linearly independent generators L_{ij} of symmetries of unispace, when formulated as expressions in the x_j' and $\partial/\partial x_j'$, differ from the 11 generators of the group of global conformal transformations in Minkowski space by terms of order $1/R^2$, as $R \to \infty$.

Remark 1 It might appear anomalous that 15 vector fields converge to 11 vector fields. What happens is that two ordered sets, each consisting of four of the L_{ij}, converge to the same conformal vector fields in Minkowski space.

Remark 2 This result is independent, as are many in this chapter, of the dimensionality of space–time.

Proof We take the Minkowski energy–momentum vector in the usual form

$$\left(-i\frac{\partial}{\partial t'}, i\frac{\partial}{\partial x'}, i\frac{\partial}{\partial y'}, i\frac{\partial}{\partial z'}\right),$$

and define the (physically scaled) uni-energy–momentum vector to have the form

$$(-iR^{-1}L_{-1,0}, -iR^{-1}L_{-1,1}, iR^{-1}L_{-1,2}, iR^{-1}L_{-1,3}).$$

From earlier obtained expressions for the L_{ij}, it follows that:

(the univector) – (the special relativistic vector)

= (inverted relativistic vector),

where the latter is defined as the transform of the special relativistic energy–momentum vector through conformal inversion. That is, the vector

$$-iQ\left(\frac{\partial}{\partial t'}, -\frac{\partial}{\partial x'}, -\frac{\partial}{\partial y'}, -\frac{\partial}{\partial z'}\right)Q$$

$$= \frac{-L_{-1,0} + L_{0,4}, L_{-1,1} - L_{1,4}, L_{-1,2} - L_{2,4}, L_{-1,3} - L_{3,4}}{-i(2R)}$$

The jth component of this vector is

$$\pm i\left[\frac{t'^2 - x'^2 - y'^2 - z'^2}{4R^2}\frac{\partial}{\partial x'_j} - \frac{x'_j \varepsilon_j}{2R^2}\left(t'\frac{\partial}{\partial t'} + x'\frac{\partial}{\partial x'} + y'\frac{\partial}{\partial y'} + z'\frac{\partial}{\partial z'}\right)\right],$$

which is $O(R^{-2})$.

The angular momenta L_{ij} $(i, j = 1, 2, 3)$ have the form

$$-i\left(x_j\frac{\partial}{\partial x_i} - x_i\frac{\partial}{\partial x_j}\right) = -i\left(x'_j\frac{\partial}{\partial x'_i} - x'_i\frac{\partial}{\partial x'_j}\right),$$

which is independent of R, and has the same expression both in Minkowski and unispace. The same is true of the boosts

$$-iL_{0,j} = -i\left(x'_0\frac{\partial}{\partial x'_j} + x'_j\frac{\partial}{\partial x'_0}\right) \qquad (j = 1, 2, 3),$$

and the infinitesimal scale transformation $K = \sum_j x'_j\,\partial/\partial x'_j$.

In particular, $R^{-1}L_{-1,j}$ and $R^{-1}L_{j,4}$ both differ from $\varepsilon_j\,\partial/\partial x'_j$ by $O(R^{-2})$, and agree in the limit $R \to \infty$ with the conventional energy–momentum component $\varepsilon_j \partial/\partial x'_j$. The differences $L_{-1,j} - L_{j,4}$ thus are locally approximate absolute constants of the motion. As such they are locally approximately representable by a slowly-varying vector field, which physically would appear most naturally as potentially related to gravitational phenomena as in the Weyl–Veblen theory, but possibly related also to microscopic processes as internal quantum numbers. There is no clear connection with extragalactic observation, and these generators will not be treated further here.

The philosophy of the chronometric approach to elementary particle theory may be briefly indicated here, as a means of clarifying its coherence with both macro- and microphysics. Its basic premise is that while the cosmos as a whole is covariant with respect to $SU(2, 2)$, or more precisely its universal covering group, say G_{15}, the observable microcosmos is covariant only with respect to the scale-covariant subgroup, say G_{11}. The scale generator $-L_{-1,4}$ is, like the superrelativistic energy-momentum vector, locally an approximate absolute constant of the motion and thus determines a slowly-varying scalar field. This again is most naturally interpreted from a gravitational standpoint, in terms, for example, of the Nordstrom-like theory or, in combination with the vector field just indicated, the Weyl–Veblen theory. From an elementary particle standpoint, what may be important is the restriction of the G_{11} to the usual G_{10} Lorentz group brought about by the elimination of the scale generator; thereby conformal invariance is not at all inconsistent with the existence of massive particles. The G_{11} is invariantly specified as the subgroup of $\widetilde{SU}(2, 2)$ leaving invariant the "infinite" points relative to the observer in question. These are the points that are carried by the covering transformation from unispace $\tilde{M}$ to the conformal compactification $\bar{M}$ of Minkowski space M into points that are in $\bar{M}$ but not in M; i.e., the antipodal point, all points lightlike relative to it, and all transforms of these points by the center of G_{15}. An alternative mechanism for the introduction of massive particles is the restriction to the $SO(2, 3)$ subgroup, which includes time development, space rotations, etc., i.e., that consisting of transformations leaving unaltered the last coordinate ξ_4. This mechanism would probably lead to a countably infinite series of theoretical particles of discrete masses without the interventation of adjustment for scaling, which could conceivably prove to be unobservably small.

While these ideas are qualitative, they are nevertheless suggestive of relatively concrete models for the basic elementary particles. The leptons, for example, may most simply and naturally be correlated with the solution space of the Dirac equation in $S^1 \times S^3$, or in $U(2)$. The Z_4 central subgroup of $SU(2, 2)$ provides a quantum number that can naturally be expected to distinguish neutral from charged leptons. The vanishing of the parameter m in the Dirac equation does not imply the vanishing of the physical masses of all the elementary particles involved here (i.e., irreducible constituents of the indicated representation of $SU(2, 2)$ on its restriction to the extended Poincaré subgroup), due to the curvature of space and to the role of scaling.

Similarly the baryons are quite conceivably represented by the totality of spinor fields on $\tilde{M}$, or one of its locally isomorphic versions, having "real mass," i.e., the square of the Dirac operator has a nonnegative spectrum in the state corresponding to the field (more exactly, semibounded spectrum, since the curvature of space displaces the zero-point). On restriction to G_{11}, the representation of the G_{15} defined by these fields may well split into only a finite number of irreducible constituents. (I am indebted to B. Kostant for citations of analogous known group-representation-theoretic phenomena.) The scale then becomes an experimental constant, but the ratios of the masses of the constituents would be mathematically computable.

These computations are technically fairly advanced, but appear entirely feasible as a program for immediate development. Of course, it is always possible that nonlinear interaction effects are so large as to dominate completely the spectrum of apparent elementary particles and to reduce the implications of the present group-theoretic approach to a qualitative level. This is, e.g., the present position of the Heisenberg school, among others. However, the discreteness and unicity of the leptons and baryons are strongly suggestive that elementarity of particles is at least partially real rather than relative; and the mechanisms here proposed for the classification of these particles are considerably more unique analytically and definitively accessible quantitatively than those which depend on proposed nonlinear quantized fields. In addition, the—possibly surprising—usefulness in cosmology of rational methods, based on general considerations of causality and symmetry, etc., evidenced by the good agreement with observation of the chronometric theory, must be taken into consideration, however different physically the situations in elementary particle physics and extragalactic astronomy may appear to be.

IV

Astronomical applications

1. Introduction

There are logically two parts to the astronomical applications: (A) the elucidation of the theoretical implications for observable quantities; (B) the comparison of the theoretical predictions with actual observations.

Part A is contained in the following three sections of this chapter; these explore the following theoretical relationships, from the standpoint of the present theory and its comparison with the expanding-universe theory:

(1) the redshift–magnitude relation for a single luminosity class;

(2) the redshift distribution for a single luminosity class of uniform spatial distribution;

(3) the Schmidt luminosity–volume ratio for apparent-luminosity-limited samples of uniform spatial distribution;

(4) the magnitude–aperture relation for galaxies, and its implications for cosmological tests based on (1) and (2) above;

(5) the $\log N - \log S$ relation for radio sources of given luminosity function and spectral index;

(6) the metric angular diameter–redshift relation for a uniform class of objects uniformly distributed in space.

While some further independent tests may be envisaged, the data presently available are for the most part totally inadequate for statistically significant studies.

The theoretical treatment is illustrated in Part A by reference to actual observations; however, the detailed description of the comparison between the best available data and the predictions of Part A is given in Sections 5–20 constituting Part B. There are separate sections on galaxies and quasars. Emphasis is on statistically controlled data and the utilization of standard statistical tests; however, extensive runs of data, even when of uncertain statistical homogeneity, are also discussed. In the last section, predictions and observations are compared for cases in which statistical levels of significance are uncertain and theoretical values, such as intrinsic simplicity, or economy in the use of parameters or energy, are involved in the comparisons. Examples of such cases include: the apparent near cutoff in the number of quasars above redshift 3; the energy output of quasars in comparison with that of galaxies; apparent superlight velocities; apparent rarity of quasars in identifications of optically very faint radio sources.

2. The redshift–magnitude relation

With the notation of Chapter III, consider a luminous object at a distance ρ in natural units from the point O of observation, which we take as origin. The redshift z is then $\tan^2(\rho/2)$, according to Chapter III. By the inverse square law of luminosity decrease with distance, and spherical geometry, the luminosity I, as a function of ρ, varies as $(\sin \rho)^{-2}$, apart from redshift effects. If at the source the spectral function is $f(\nu)$ and if observation is made in a frequency range $\nu_1 < \nu < \nu_2$, then the energy at the source contributing to this range is $\int_{\nu_1(1+z)}^{\nu_2(1+z)} f(\nu)\, d\nu$. This energy is not only diffused according to the inverse square law but also redshifted, or diminished by the factor $1 + z$ as observed. Consequently the observed luminosity L_{obs} varies as

$$\frac{1}{\sin^2 \rho} \frac{1}{1 + z} \int_{\nu_1(1+z)}^{\nu_2(1+z)} f(\nu)\, d\nu.$$

Assuming that $f(\nu) \propto 1/\nu^\alpha$, it follows from this equation and trigonometry that

$$L_{\text{obs}} \propto \frac{(1 + z)^{2-\alpha}}{z},$$

or in term of magnitudes,

$$m = 2.5 \log z - 2.5(2 - \alpha) \log(1 + z) + C.$$

For $\alpha \geqq 1$, m is then a monotone increasing function of z, which if $\alpha = 1$ attains the finite limiting value C as $z \to \infty$. It attains this limiting value relatively rapidly; at $z = \frac{2}{3}$, the brightness is within one magnitude of the limiting value. The finiteness of this limit is of course purely theoretical; any actual physical source, having finite energy, cannot have a constant spatial index $\alpha \leqq 1$, since otherwise the integral $\int^{\infty} f(v)\,dv$ representing the total energy would be divergent; indeed, in actuality $f(v) = 0$ for sufficiently large v. However, at present, there is little evidence to suggest that variation in the spectral index is a significant factor over the presently observable redshift range, as regards the m–z relation for quasars, and as in most treatments of the m–z relation it will be neglected.

In any event, over a redshift range such as $0.4 < z < 2.5$, the chronometric theory predicts a dimming in apparent magnitude of < 1 mag for objects such as quasars of spectral index ~ 1, while the Hubble model† (or Friedmann models having reasonable parameters) predicts a dimming of ~ 4 mag. Thus, barring extreme observational difficulties with the larger redshift ranges, the difference between the chronometric and expansion theories redshift predictions should not be difficult to detect. It will be found that, indeed, quasar observations are in quite satisfactory agreement with the chronometric prediction, but reject the Hubble law, at a high level of certainty. The latter result may be regarded as a form of demonstration of very strong evolutionary effects, within the expansion theory framework, as will be discussed later.

The theories also differ markedly for small z; according to the chronometric theory,

$$m \sim 2.5 \log z + \text{const}, \qquad 0 < z \leqq 0.1.$$

There is thus a difference of 2.5 mag between the chronometric and expansion theories' predictions over either of the ranges, $0.001 \leqq z \leqq 0.01$ and $0.01 \leqq z < 0.1$. Unfortunately, there are observational difficulties in these ranges, which may be less significant for quasar observations; these have been stressed for a long time by workers in the field. They are primarily: (a) the relatively greater difficulty of observing "standard candles" (in particular, "selection effects"); (b) the aperture effect; (c) the intrinsic velocities of the luminous objects; and, secondarily, (d) the "K-effect," and (e) galactic absorption, which is relatively small for the magnitude differences

† For brevity and in conformity with general usage, the term Hubble model (or theory) is used in the present work to indicate the Doppler model in which space is Euclidean and the redshift is proportional to distance. Historically, however, the term is a considerable oversimplification. On several occasions, partly with collaborators, Hubble expressed clear reservations about the Doppler theory of the redshift, but stated the opinion that the only likely theoretical alternative was a new fundamental physical development. Moreover, the redshift–distance relation was initially reported as "roughly linear" and later as involving additionally a definitely positive quadratic term; see Hubble (1936a).

involved. (Correction for galactic evolution is required only in the expansion theory.) As emphasized by Humason *et al.* (1956), the aperture effect is strongly z-dependent, and must be properly corrected for, in order to have a valid basis for comparison between theory and observation. On the other hand, as is clear from recent work of Sandage (1972a), the aperture correction is in practice model-dependent. As a consequence, for objects and redshift ranges in which the aperture effect is significant, the redshift–magnitude relation must be of a quite detailed nature as regards the surface brightness profile of galaxies, etc., in order to be testable. The data on bright cluster galaxies are such that each theory fits part of it well and part of it equivocally.†

Intrinsic ("peculiar") velocities are a conceivable difficulty in dealing with very small redshifts. In principle, the difficulty could be overcome if sufficient randomized data are available. However, there is no model-independent indication that the problem is a serious one, apart from the motion of the Sun and Galaxy. The catalog of the de Vaucouleurs (1964) includes only 14 blue-shifted objects, with an average blue shift < 100 km/sec, among more than 740 for which redshifts and magnitudes are given.

In any event, the existing galaxy data at low redshifts are in poor agreement with the Hubble law. A lengthy study by G. de Vaucouleurs (1972) led him to postulate a local spatial anisotropy. This has been disputed by Sandage *et al.* (1972). The chronometric theory is in excellent agreement with the data, its prediction for the m–z relation being substantially the empirical law $m = 2.5 \log z + \text{const}$ found by de Vaucouleurs. This law is confirmed by maximum-likelihood estimation, whether for the totality of 742 redshifted galaxies with m–z–θ data listed on the de Vaucouleurs tape (updated to 1972) of their guide (1964), or for subsamples selected on morphology, field of the sky, or both. It is in addition in distinctly better agreement with the classic data of Humason, Mayall, and Sandage than is the Hubble Law, as first indicated by Hawkins (1962), in the case of field galaxies.

Closely related to the redshift–magnitude relation is the redshift distribution law. Assuming a uniform spatial distribution of luminous objects, i.e., that the number of objects of a given type in a given region of space is proportional to the volume of the region, the fraction of objects out to a

† Quasi-phenomenologically (see below), the data of Sandage on the m–z relation (for a sample derived from 41 clusters) is in considerably better agreement with the Hubble law than the chronometric prediction; but the latter does effect a 45% reduction in dispersion. On the other hand, the compilation by Noonan of all published redshifts for clusters (146 in all) is in much better agreement with the chronometric than the Hubble prediction regarding the $N(< z)$ relation.

given redshift z is well determined in both theories, and well known in the expansion theory; for small z, this fraction $F(z)$ varies approximately as z^3, independently of the precise parameters of the theory (or exactly as z^3 in the original Hubble theory based on Euclidean space), and increases rapidly with increasing z in the observational range. To treat the matter in the chronometric theory, let $V(\rho)$ denoted the volume of space up to distance ρ from the origin O; then dV varies as $\sin^2 \rho \, d\rho$. Integrating, it follows that

$$V(\rho) = \tfrac{1}{2}(\rho - \sin \rho \cos \rho) = [\tan^{-1} z^{1/2} - z^{1/2}(1 - z)(1 + z)^{-2}],$$

expressing ρ in terms of z. This implies that $F(z)$ varies as $z^{3/2}$ for small z, in the chronometric theory, as was to be expected from the approximate quadratic dependence of z on distance for small distances. This represents a considerable difference from the behavior indicated by the expansion theory. The existing samples which are or may be free from serious selection effects are limited in number and size, but they favor the $\frac{3}{2}$ power law over the third-power law, whether galaxies or quasars are used.

When selection by luminosity is an important factor, the V/V_{m} test treated by Schmidt (1968) may still be used, provided the sample is complete out to a definite limiting magnitude. If this limiting magnitude is $\bar{m}$, then the ratio of the volumes $V(z)/V(\bar{z})$, where $\bar{z}$ is the maximum redshift for which the object would remain in the sample (if located at the distance indicated by the redshift), and $V(z)$ denotes the volume of the region in space in which the redshift is bounded by z, is uniformly distributed in the interval [0, 1], on the hypothesis that the luminous objects in question are uniformly distributed in space. In the Hubble theory, e.g., in its simplest form, $\bar{m} - m = 5 \log \bar{z} - 5 \log z$, whence V/V_{m} (setting $V(z) = V$ and $V(\bar{z}) = V_{\mathrm{m}}$, to conform with the notation of Schmidt) takes the value $10^{-0.6(\bar{m}-m)}$.

In the chronometric theory, for objects of spectral index ~ 1

$$\bar{m} - m = 2.5 \log[\bar{z}/(\bar{z} + 1)] - 2.5 \log[z/(z + 1)];$$

a simple computation leads to the result

$$\bar{z} = \gamma/(1 - \gamma), \qquad \gamma = [z/(z + 1)]10^{0.4(\bar{m}-m)};$$

if $\gamma > 1$, then the luminosity is so great that the object would be included in the sample anywhere in the space under consideration. Setting

$$\rho = 2 \arctan z^{1/2}, \qquad \bar{\rho} = 2 \arctan \bar{z}^{1/2},$$

$$\frac{V}{V_{\mathrm{m}}} = \left(\rho - \frac{\sin 2\rho}{2}\right) \bigg/ \left(\bar{\rho} - \frac{\sin 2\bar{\rho}}{2}\right).$$

The given expression for V/V_{m} differs by little from the ratio $(z/\bar{z})^{3/2}$ in the indicated redshift range, being typically of the order of 1% greater. As $z \to 0$, this expression converges to $10^{-0.6(\bar{m}-m)}$, i.e., to the expansion-theoretic value, but for moderate values of z, the difference between the respective values of V/V_{m} may be significant. Even for small values of z, the test is potentially discriminatory on the basis of the model dependence of the magnitudes, via the corresponding dependence of the appropriate aperture.

The distribution laws differ markedly for large z, in as much as $V_{\mathrm{chrono}}(z)$ attains a finite limiting value which is approached for much lower z than in the case of realistic closed Friedmann models. For example, $N(2.25 < z < 3)/N(z < 2.25)$ has the value 1.37 in the (nonevolutionary) Hubble theory, > 0.33 for the Friedmann models with $q_0 \leqq 1$, and the value 0.09 in the chronometric theory (in which the region $2.25 < z < 3$ corresponds to the zone in space where the polar angle ρ lies in the narrow range $112.6° < \rho < 120°$). The latter value is greater than, but in the light of the observational situation agrees within an acceptable level of random fluctuations, with the value observed for quasars; however (indeed, virtually as a consequence), the former do not, even allowing for various effects which may modify the value (absorption, K-effect, etc.).

To summarize, the redshift–magnitude relation, and the related redshift distribution, provide several quite disparate theoretical predictions of a direct and straightforward nature. Their confrontation with observation may therefore reasonably be expected to furnish counterindications for at least one of the theories, although not necessarily positive indications for the other.

3. Further cosmological tests

We attempt no exhaustive analysis, but treat only two additional tests which may fairly soon become statistically applicable as data improve.

a. The redshift–angular diameter relation

It must be emphasized that the theory treats the *metric* angular, rather than isophotal angular diameter. For the relation of the latter diameter to the redshift is quite complex and dependent on a variety of uncertain functions. On the other hand, the metric diameter is in general not directly observed, so that the relation derived is not readily checked against observation.

In the chronometric theory, an object of metric diameter d has at distance ρ the angular diameter $\theta = d/\sin \rho + O(d^2)$, employing the same natural units as earlier. For galaxies, d^2 is negligible, and employing the redshift–diameter relation, it follows that

$$\theta = d(1 + z)/2z^{1/2}.$$

The chronometric θ–z relation appears quite different from the Hubble relation $\theta \propto z^{-1}$, and from the relation treated by Sandage (1972a) $\theta \propto (1 + z)^2/z$. Unfortunately, in the z range in which there are measurements that can reasonably be construed as metric diameters, the various theoretical diameters differ by far less than the intrinsic dispersion in the angular diameters. The main difficulty in the use of the θ–z relation is indeed that of determining the angular diameter of a more-or-less constant metric diameter. The isophotal and metric diameters in general may behave differently as functions of redshift.

The very large sample represented by the de Vaucouleurs tape gives $\theta \propto z^{-1/2}$ to a much better approximation than $\theta \propto z^{-1}$. As in the case of the m–z relation, the dispersion from the expansion prediction is always of the same order of magnitude as the dispersion in the apparent quantities, while the dispersion from the chronometric prediction is materially less, being generally of the same order of magnitude as that from the least-squares fit. Again, this is true for the entire sample with appropriate data, or for subsamples selected on morphology, field of sky, redshift interval, etc. Also, here, as in the case of the m–z relation, proper analytical allowance for the observational cutoff in apparent magnitude, i.e., for the conceivable material z-dependence of the sample arising from the possible existence of significant numbers of intrinsically bright but apparently faint galaxies which have been excluded from the sample, merely slightly improves the fit of the chronometric relation.

The much smaller sample of brightest cluster galaxies treated by Sandage (1972a) is represented by him as following an approximate Hubble law, $\theta \propto z^{-1}$. However, quite apart from the apparent selection effects previously noted in connection with this sample, the results in the θ–z analysis are quite sensitive to the inclusion or exclusion of the objects at extreme and isolated redshift ranges. For example, when the local region $cz < 4000$ is excluded, and also those beyond the gap of > 2000 in the sample values of cz near $cz = 18{,}000$, in the sample of galaxies whose isophotal diameters were estimated from 48-inch Schmidt plates, there is little difference between the fits of the $\theta \propto z^{-1}$ and $\theta \propto z^{-1/2}$ curves. This subsample of thirty-five galaxies, defined by restriction to the range $4000 < cz < 18{,}000$, constitutes the major and most coherent portion of the cited sample. The dispersions in the deviations of $\log \theta$ from the $-\log z + \text{const}$ and $-0.5 \log z + \text{const}$ lines are, respectively, 0.115 and 0.119. The sample of nineteen galaxies measured from 200-inch plates is quite irregular in its redshift distribution and devoid of published statistically viable selection criteria; its statistical weight compared to the vastly larger BGC sample and the various subsamples indicated appears consequently to be quite small.

The systematic observations by Baum (1972) of galaxy diameters are nevertheless statistically too limited, as well perhaps as too complex in theoretical interpretation, to differentiate between the relevant θ–z relations. The compilations of double radio source angular diameters by Legg (1970) provide diameters which are probably substantially metric, but include quite heterogeneous data, and are too limited in sample size for definite statistical conclusions to be drawn. If one computes the discriminatory variance of the linear diameters in Legg's data, say in kiloparsecs (kpc), it is found that it is substantially smaller in the chronometric theory than in a typical expanding-universe theory. However, this is a result of the smaller overall distance scale of the chronometric model, based on the value $H \sim 80$ at 10 Mpc; the logarithms of the linear diameters in the two theories have variances of the same order of magnitude. The compilation of radio angular diameter data by Miley (1971) is likewise too heterogeneous and/or limited in sample size to be statistically discriminatory. The variation of apparent angular diameter with frequency, and its dependence on spectral index, are further obstructions to the use of radio angular diameters without much more data and analysis. But the θ–z observations are effective in indicating a nontrivial dependence of θ on z for larger z, and the corollary correlation of z with distance.

b. *The* log N–log S *relation*

At the present time, uncertainties as to luminosity and spectral functions, as well as observational ambiguities involving faint sources, limit the precision of this test. However, there appears to be a substantial qualitative difference in this respect also between the chronometric theory and nonevolutionary expansion theories. In the latter theories, it seems quite difficult to obtain values of the index $\beta = -\partial \log N/\partial \log S$ which are greater than the Euclidean value 1.5, with what appear as a priori reasonable choices for spectral and luminosity functions; cf. e.g., Longair and Rees (1972). In the chronometric theory, such larger values are readily attained. Moreover, with simple reasonable models for the luminosity function and choice of spectral index, a $N(S)/N_0(S)$ curve is obtained which shows the key observed qualitative features of the observational N–S relation (cf. Longair and Rees (1972); here $N_0(S)$ denotes the corresponding function in an Euclidean universe, i.e., $N_0(S) \propto S^{-3/2}$).

Consider, to begin, with a uniformly distributed class of objects of fixed luminosity L, and having spectral index α. Let z denote the redshift of a source and ρ the corresponding distance in natural units: $\rho = 2 \tan^{-1} z^{1/2}$. In the chronometric theory, the observed luminosity is proportional to $(1 + z)^{2-\alpha} z^{-1}$ as earlier derived; the expected number $N_L(S)$ of sources apparently brighter than S, within the luminosity class under consideration, is then proportional to the volume of space within which the indicated function of z (or equivalently, function of ρ) is greater than S.

It follows that

$$N(S) \propto \int_{L_{obs}(\rho) \geq S} \sin^2 \rho \, d\rho,$$

where $L_{obs}(\rho) = L(1 + z)^{2-\alpha} z^{-1}$, expressed as a function of ρ. The behavior of $N(S)$ and the corresponding population-brightness index $\beta = -\partial \log N / \partial \log S$ depends significantly on α, and especially on whether $\alpha \geqq 1$ or $\alpha \leqq 1$. If $\alpha \geqq 1$, $L_{obs}(\rho)$ is a monotone decreasing function of ρ, and, for $\alpha = 1$, it follows that

$$N(S) \propto (\rho - \sin \rho \cos \rho), \qquad \rho = \cos^{-1}(1 - 2(L/S));$$

$$\beta = \frac{2 \sin \rho (1 - \cos \rho)}{\rho - \sin \rho \cos \rho}.$$

If $\alpha < 1$, $L_{obs}(\rho)$ decreases down to a certain minimum at $z = 1/(1 - \alpha)$, and then increases again as ρ (or z) continues to increase. It follows that

$$N(S) \propto \int_0^{\rho_1} \sin^2 \rho \, d\rho + \int_{\rho_2}^{\pi} \sin^2 \rho \, d\rho,$$

where ρ_1 and ρ_2 are determined by the equation

$$LS^{-1} = 2^{\alpha-2}(1 - \cos \rho)(1 + \cos \rho)^{1-\alpha},$$

and the inequalities

$$0 < \rho_1 \leqq \rho_2 < \pi.$$

Here L is a parameter proportional to the intrinsic luminosity of the source. It results from a simple computation that

$$\beta = 2\left(\frac{\sin^3 \rho_1}{\alpha + (2 - \alpha) \cos \rho_1} - \frac{\sin^3 \rho_2}{\alpha + (2 - \alpha) \cos \rho_2}\right)$$

$$\times \left(\rho_1 - \sin \rho_1 \cos \rho_1 + \pi - \rho_2 + \sin \rho_2 \cos \rho_2\right)^{-1}.$$

It follows that β becomes infinite as the source strength decreases to that for which $\rho_1 = \rho_2$, i.e., for which the source is observable anywhere in the universe. For smaller S, $\beta \equiv 0$, since N can become no larger. The situation is well represented by the simple case $\alpha = 0$, which gives

$$\beta = \sin^3 \rho (\cos \rho)^{-1} (\rho - \sin \rho \cos \rho)^{-1}; \qquad \rho = \sin^{-1}(L/S)^{1/2}.$$

The situation for $0 < \alpha < 1$ is qualitatively obtainable by interpolation between the values $\alpha = 0$ and $\alpha = 1$. Typically, as S decreases from high values, β begins eventually to increase perceptibly and rises eventually to ∞; but for

TABLE 1
The population-brightness index $\beta = -\partial \log N/\partial \log S$ for sources of fixed intrinsic luminosity and spectral index α in the chronometric theory

	β		
log S	$\alpha = 0$	$\alpha = 0.7$	$\alpha = 1$
2.0	1.503	1.501	1.499
1.8	1.505	1.501	1.498
1.6	1.508	1.502	1.496
1.4	1.512	1.503	1.494
1.2	1.520	1.505	1.490
1.0	1.533	1.508	1.485
0.8	1.554	1.515	1.475
0.6	1.594	1.528	1.459
0.4	1.675	1.563	1.432
0.2	1.895	1.698	1.379
0.0	∞	∞	1.275
<0	0	0	0

fainter values of S it is identically zero. Table 1 shows the values of β for the cases $\alpha = 0$, 0.7, and 1, and can be used to estimate β roughly for other values of α by interpolation.

The discontinuous behavior near $S = L$ for $0 \leqq \alpha < 1$ is smoothed out by a smooth luminosity function, but the same qualitative behavior is otherwise manifested. If $P(L)$ is the relative number of sources of intrinsic luminosity less than L, the resultant $N(S)$ takes the form $\int G_L(S)\, dP(L)$; if the spectral index is permitted to vary, there will in addition be a corresponding integral over its range. In presenting quantitative results, it is convenient to follow the practice of dealing with the ratio $N(S)/N_0(S)$, where $N_0(S)$ denotes the corresponding Euclidean quantity, and so is proportional to $S^{-3/2}$; the proportionality factor may conveniently be chosen so that $N/N_0 \sim 1$ for large S. The resulting expression is

$$N/N_0 = \tfrac{3}{2}\left(\int G_L(S)\, dP(L)\right) \Big/ \int L^{3/2}\, dP(L) S^{3/2}.$$

The luminosity function for radio sources is not well determined, but is thought (expansion theoretically) to be rather broad. In Figure 4 the N/N_0 curve for a single luminosity class having spectral index 0 has been plotted, together with the smoothed-out curve resulting from a hypothetical lumino-

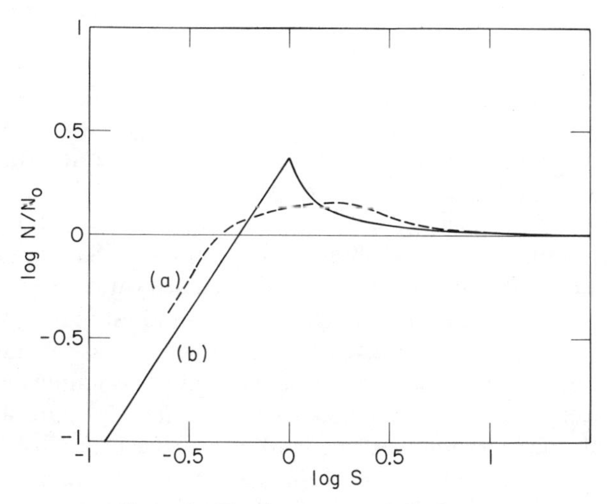

Figure 4 *The chronometric N–S relation.*

Curve (a), ~ 1 decade range in luminosity; curve (b), single luminosity class. Assumptions: (1) spatially uniform distribution of sources; (2) spectral index $= 0$. The rise above the horizontal axis corresponds to values of $-\partial \log N/\partial \log S$ in excess of the Euclidean value 1.5, and takes place for all values of the spectral index < 1.

sity function corresponding to a range of 0.9 in log L, log L being uniformly distributed in this range. A change in the L-scale merely translates the curve horizontally. The curve is qualitatively similar to observational curves (cf., e.g., Longair and Rees, 1972), showing the key features of a rise in population-luminosity index above $\frac{3}{2}$ as the source strength declines from very strong to strong, followed by a decline, eventually falling well below the Euclidean value for faint sources. Averaging over the spectral index, typically ~ 0.3 for flat sources or 0.7 for steep ones, but fairly broadly dispersed, will reduce the qualitative effects indicated, while the use of a less broad luminosity function would increase them. For $\alpha = 1$, the curve differs by little over the physically relevant range from the constant value unity; for very low flux levels, β falls to ~ 1.275, before vanishing identically.

Besides the surely relevant and poorly known luminosity function and distribution of spectral indices, further factors may be relevant. Definitive statistically testing on the basis of the N–S relation will not be highly discriminatory until such matters are settled, and difficulties and discrepancies in the reported observations of faint sources are resolved. However, there is no apparent reason to anticipate that the various effects involved are of magnitudes sufficient to alter the disagreement of the observed N–S relation with a nonevolutionary Friedmann cosmology, or its agreement within statistical fluctuation with the present theory (cf. below).

4. The aperture correction for galaxies

It has long been recognized that the aperture correction to galaxy magnitudes is a matter of difficulty and delicacy, and yet at the same time of considerable importance to the redshift–magnitude relation. In particular, Humason *et al.* (1956) in their classic paper emphasized the highly material z-dependence of the aperture correction, and the necessity of compensating for it if valid results were to be obtained. In recent work Sandage (1972a) has again treated the aperture correction, and discussed within the framework of general relativistic models the practical problem of estimating the luminosity of portions of galaxies of fixed metric, rather than isophotal, diameter.

Unfortunately, no model-independent means of obtaining galaxy magnitudes for central portions of a fixed metric diameter is established. Within the limits of Friedmann models, a complex recursive procedure is indicated by Sandage, but demonstrations of the convergence of the method and the unicity of its results are lacking. Within the much broader limits encompassing general relativistic models, the present chronometric model, and others of comparable nature, it seems hopeless to seek a unique result.

This suggests that the magnitude–redshift relation for galaxies should primarily be employed as a means of testing hypotheses. In particular a $\partial m/\partial \log z = 5$ apparent slope for large-aperture measurements of "standard candles" is not entirely a simple observational fact, but in significant part a theoretical inference which is in agreement with observations, when the observations are made and reduced in accordance with the theory in question. In principle, it is possible that a different value for $\partial m/\partial \log z$ may be equally valid, from the standpoint of essentially the same observations, but a different theory. Indeed, this possibility is well exemplified by consideration of a conceivable attempt to avoid aperture corrections by observing only very narrow central portions of galaxies. Quite apart from the rapid decrease in surface brightness just beyond the center which might well obviate such an approach, it is demonstrable on theoretical grounds that such measurements would be incapable of discriminating at low redshifts (say less than 0.1) between the redshift laws $z \propto d^\alpha$ for a wide range of exponents α.

It will suffice to contrast the Hubble law $z \propto d$ (say $z = c_1 d$) and the Lundmark law $z \propto d^2$ (say $z = c_2 d^2$; Lundmark (1925) fitted a quadratic polynomial, but his name may serve appropriately). These are here to be regarded as hypotheses, to be tested by observations on the same galaxies, forming "standard candles," of substantially constant absolute luminosity and surface brightness characteristics. The apertures must be adjusted in accordance with the respective hypotheses, to obtain the luminosity of central portions of each galaxy of constant metric diameter characteristics.

Consequently the apertures must be adjusted in accordance with the respective hypotheses, to obtain the luminosity of the central portion of each galaxy of a fixed metric diameter. If a portion of metric radius r is to be observed, the respective apertures then vary as a/z and $a/z^{1/2}$, respectively. The resulting situation can then be summarized as the

Theorem If *either* one of the Hubble or Lundmark laws is valid, then *both* sets of observations—at the respective appropriate apertures—will be in agreement with *both* corresponding laws.

Proof Let $L(\theta)$ denote the luminosity of a given galaxy as observed with aperture θ; let $I(r)$ denote the surface brightness of the galaxy as a function of the distance r from the center. Then $L(\theta) \propto a^{-2} \int_0^{\theta a} I(r)r\,dr$, but for small apertures, $\int_0^{\theta a} I(r)r\,dr \sim I(0)(\theta a)^2/2$. It results that $L(\theta) \propto \theta^2$, implying that the corresponding magnitude $m(\theta) = -5 \log \theta + k$, k being a constant. Agreement between observations and the Hubble law means that if $\theta_1(z)$ denotes the appropriate aperture on the basis of the Hubble theory, then $m(\theta_1(z)) = 5 \log z + k'$, k' also being a constant. Agreement between observations and the Lundmark law similarly means that $m(\theta_2(z)) = 2.5 \log z + k''$, where $\theta_2(z)$ is the appropriate aperture on the basis of the Lundmark law. Since $\theta_1(z) \propto z^{-1}$ on the basis of the Hubble law, and $\theta_2(z) \propto z^{-1/2}$ for the Lundmark law, $m(\theta_1(z)) - m(\theta_2(z)) = 2.5 \log z$, which is precisely the difference in magnitudes which would be observed.

Thus there is not even a theoretical possibility of using observations at small apertures to discriminate between the Hubble and Lundmark redshift–magnitude laws for galaxies. On the other hand, at large apertures, contamination from stars, the brightness of the night sky, etc. become serious limitations. There remains a possibility that observations at intermediate apertures may sufficiently avoid both problems; but the only published data that appear to be statistically applicable to the question do not substantiate the possibility.

The bright cluster galaxy observations of Peterson (1970a), taken at apertures $\theta_1(z)$, can be corrected by a standard curve to the apertures $\theta_2(z)$; these corrections are relatively crude in that there is quite considerable variation between the surface-brightness curves $I(r)$ for such galaxies. Nevertheless, no statistically significant difference between the fits to these data of the Hubble and Lundmark laws is apparent. It should be of interest to make direct measurements of the magnitude of the Peterson galaxies at the apertures indicated by the chronometric theory, and to make measurements of additional galaxies, chosen in a statistically controlled manner, at both apertures $\theta_1(z)$ and $\theta_2(z)$. It should be borne in mind, however, that these apertures depend also on the Hubble parameter (itself z-dependent in the chronometric theory); therefore, measurements at several corresponding apertures should be taken.

The apparent limitations on the redshift–magnitude relation for galaxies in discriminating between the two laws means neither that galaxy observations as a whole are without discriminatory potential, nor that the redshift–magnitude relation is inherently ineffective. Indeed, the Schmidt V/V_{m} test is applicable to samples of galaxies which are complete to fixed apparent magnitudes, and has significant results for the Peterson sample: it is spatially extremely nonuniform according to this test, within the expanding-universe framework, but does not deviate significantly from spatial uniformity according to the chronometric theory. This suggests that apparent local superclustering emphasized by G. de Vaucouleurs (1970), following relevant observations of Holmberg, may not necessarily be physically real, but quite possibly largely a consequence of the theoretical framework within which the observations are analyzed. Observations in other fields which are complete out to fixed apparent magnitudes, or selected from complete lists in a statistically random fashion, could be used both for a definitive check on the redshift–magnitude relation, as indicated, and to test overall spatial uniformity. In any event, the redshift–magnitude relation for quasars is useful for discriminating between the expansion and chronometric theories; not only are aperture corrections not required, but their qualitative implications at larger redshifts are entirely different.

Consider now the practical problem of estimating the observed magnitude of a given galaxy at aperture θ', given the observed magnitude at aperture θ. Most applicable methods take the surface brightness at distance r from the center of the galaxy, normal to the line of sight, to have the form $I(r/a)$, where a is a parameter dependent on the galaxy in question; admittedly the notion of "center," and especially of distance in other than E0 galaxies must be suitably interpreted. The function I has been variously graphically presented, or taken specific analytical form. For computational purposes, the latter is more convenient; the simplest form is the Reynolds–Hubble law: $I(r) \propto (1 + r)^{-2}$, except for large r; we shall follow Abell's form for large r, i.e., $(1 + r)^{-3}$, joining it continuously to the earlier form at $r = 21.4$ as indicated by Abell and Mihalas (1966). It would make no essential difference in the following (i.e., for cosmological testing) if we used instead the form given by de Vaucouleurs, or even the Hubble form for large r as well. If the distance of the galaxy is d, then the aperture correction Δm is given by the equation

$$\Delta m = 2.5 \log[J(\theta' d/a)/J(\theta d/a)],$$

where $J(r) = \int_0^r sI(s)\, ds$. For the Abell standard form,

$$J(x) = \ln(1 + x) + (1 + x)^{-1} - 1 \qquad \text{when} \quad x \leqq 21.4,$$

$$J(x) = 22.4\left(\frac{1}{2(1+x)^2} - \frac{1}{1+x}\right) + 3.131 \qquad \text{when} \quad x > 21.4.$$

Thus the correction is determined when d and a are known, in addition to the aperture angles.

The determination of d must be made within the theoretical framework being tested, and will vary with the assumed value of the Hubble parameter. Taking, e.g., H as 100 at 15 Mpc, the chronometric and expansionary theories will give distances for the Peterson galaxies, having redshifts in the range 0.01–0.06, which differ by a factor which varies from about 0.3 to 0.7, the chronometric distance being the smaller. In Peterson's work the apertures are determined to yield in the expansionary framework fixed metric diameters of 20 kpc. Within the chronometric framework the actual metric diameters corresponding to the apertures employed are θd_c, where d_c denotes the chronometric-theoretical distance, which takes the form d_c (in kiloparsecs) $= \theta(\arctan z^{1/2})^\circ \times 0.017979$ if θ is measured in seconds, and H is as indicated (and so agrees with the value used by Peterson). Thus on the chronometric hypothesis, the observed magnitudes must be diminished by $2.5 \log[J(10/a)/J(d_c/2a)]$, to obtain the true magnitude of the central 20-kpc-diameter portion.

One thereby obtains a well-determined aperture correction, once the parameter a is specified. The determination of a again depends, however, on the theoretical model, for this determines distances, on which the conversion from angular to linear diameters depends. The chronometric a_c, for example, could in principle be determined as follows. Let m_1 and m_2 be the observed magnitudes of a galaxy at redshift z, with apertures θ_1 and θ_2, where, say, $\theta_1 < \theta_2$. Then

$$m_1 - m_2 = -2.5 \log[J(\theta_1 d/a)/J(\theta_2 d/a)],$$

where in natural units $d = 2 \tan^{-1} z^{1/2}$, the θ_j being here in radians and a in natural units. The value $a = a_c$ determined from this equation evidently depends on the assumed value of the Hubble parameter, and will differ from the value obtained by using the expansion-theoretic distance. It is therefore convenient that the aperture correction is not highly sensitive to the precise value of a, apart from a zero point correction which is irrelevant in cosmological tests. The values of a obtained in the indicated fashion actually show considerable dispersion, even within the limited class of apparent bright cluster galaxies. In the absence of systematic published work on the subject, we shall simply take $a = 1$ kpc; this may interpolate between a modal value (perhaps 0.7) and a value perhaps more likely to minimize the root mean square deviation from the true correction (perhaps 2.0), as inferred from analysis along the foregoing lines of data given by Fish (1964) and Sandage (1972b). The results of the analysis of the Peterson galaxies would not change materially if any value of a in the range 0.5–2 were employed instead. However, due to variation in a, and to the approximation for any individual galaxy involved in using a fixed surface brightness curve, an additional dispersion is introduced into the magnitudes which should eventually appear as a slightly increased dispersion in the absolute magnitudes of the galaxies, as determined from the best-fitting theoretical redshift–magnitude curve.

Since $J(10/a)$ is independent of the particular galaxy, it would suffice for cosmological testing to replace the Peterson magnitudes m by the corrected magnitudes $m' = m + 2.5 \log J(d_c/2)$. It should now be evident how the procedure may be applied to an arbitrary sample of galaxies of a specified type. These specifications must, however, be compatible with the considerations of the following section.

5. Statistical effect of the selection of the brightest objects

If one deletes from a heterogeneous list of luminous objects, quasars or galaxies, all objects fainter than a certain theoretical absolute magnitude—chronometric, expansionary, or otherwise—it has in general the effect of reducing the variance in absolute magnitude of those that are left, for all physically reasonable luminosity functions. This is the case irrespective of the validity of the theory in question, for in effect one is simply truncating a distribution beyond a certain point. The consequent reduction in variance is a statistical verity, and in no wise indicates that bright objects of the category in question form an intrinsically more homogeneous class, observation of which confirms the theory, unless the reduction is significantly greater than would arise on a statistical basis. The latter reduction is considerable, as the following analysis of a normal distribution shows. A similar analysis would apply to a mixture of normal distributions of different means; it seems unlikely that the overall figures will change greatly, for plausible types of mixtures, and we here limit the treatment to the simple cited case.

Given a zero-mean, unit-variance normal variate x, suppose the population above a value a is deleted, corresponding to the "faintest" $100p\%$ of the population. Thus the equation

$$(2\pi)^{-1/2} \int_{-\infty}^{a} \exp(-x^2/2)\, dx = 1 - p$$

gives a in relation to p. We then consider the new probability law P_a:

$$dP_a = \begin{cases} (2\pi)^{-1/2}(1-p)^{-1} \exp(-x^2/2)\, dx & \text{when } x < a, \\ 0 & \text{when } x > a. \end{cases}$$

The mean m_a of this new distribution is readily computed as $(2\pi)^{-1/2} \times (1-p)^{-1} \exp(-a^2/2)$, as is the variance in terms of the incomplete gamma function.

Table 2 gives the corresponding numerical results. Roughly speaking, deleting two-thirds of the faintest objects decreases the variance by about two-thirds. The table should be applied at each fixed redshift, as the fraction p deleted will generally vary with the redshift. The increase in dispersion for $p > 0.7$ can be understood as the effect of removing almost all of the distribution except the comparatively flat and hence widely dispersed tail.

TABLE 2
Statistical effects of selection of brightest objects

Fraction of objects deleted	Reduction in mean	Variance of remaining objects	σ_a
0.1	0.1960	0.6499	0.8061
0.2	0.3450	0.5463	0.7391
0.3	0.4967	0.4723	0.6873
0.4	0.6439	0.4194	0.6476
0.5	0.7979	0.3633	0.6208
0.6	0.9656	0.3162	0.5623
0.7	1.1590	0.3121	0.5587
0.8	1.4000	0.4185	0.6469
0.9	1.7545	0.7307	0.8548

The statistical theory has been compared with the results of selection on the absolute (theoretical) magnitudes for the Peterson sample. Deleting the faintest half of the galaxies on these bases leads to reductions in the variance of the absolute magnitudes, and in average absolute magnitudes, in quite good agreement with the statistical theory. This is equally the case whether the chronometric or the Hubble theory is employed.

6. The Peterson galaxies

Among the best data from the standpoint of statistical control are those of Peterson (1970a). These provide a complete sample of 44 bright cluster galaxies complete in a specified field to a limiting apparent magnitude of 15. The major portion of the present section is concerned with the analysis of these data along the general lines earlier indicated.

We shall later treat galaxy observations in specified categories reported by Arakelyan, de Vaucouleurs, and Sargent, among others, and discuss briefly the recently published data of Sandage. The work of de Vaucouleurs concerns nearby galaxies, and appears comprehensive and objective within reasonable statistical limits. Sargent's work contains a study of 24 Seyfert-like Markarian galaxies, characterized and observed in an apparently objective and uniform fashion. Together with the Peterson galaxies, these provide three quite different groups of galaxies. The Sandage observations overlap significantly with those of Peterson; unlike the latter, the sample that they form is not delineated in a statistically explicit fashion; for this and other reasons it does not appear possible to use them for a statistically rigorous test of the chronometric hypothesis.

Before giving the details of the analysis of the Peterson data, the central conclusions will be summarized briefly.

(a) The dispersion of the Peterson magnitudes from the best-fitting Hubble line is 0.33 mag; that from the best-fitting constant-intrinsic-luminosity chronometric curve is 0.36. The slightly greater dispersion of the chronometric theory is not statistically significant, and may well be due to the utilization of data gathered basically on the expansion hypothesis, aperture corrections being made on the basis of a fixed curve, whereas the surface brightness profiles of the galaxies do in fact vary considerably (see below regarding this question).

(b) Because of the completeness of the sample, the Schmidt luminosity–volume test is applicable. Assuming a spatially uniform population of galaxies, the ratios V/V_{m} defined by Schmidt should be uniformly distributed in the interval from 0 to 1. It is found that in actuality, they are highly skewed, and their deviation from spatial uniformity, as measured by the Kolmogorov–Smirnov statistic, is so large as to correspond to a probability of 5×10^{-5} of obtaining such a skew sample, assuming that the population is in fact spatially uniform.

(c) It is well known that it is extremely difficult to obtain a rigorously complete sample out to a given magnitude, and it is therefore conceivable that the Peterson sample is not entirely complete out to a limiting magnitude of 15, but is such to a lower magnitude, such as 14. However, a test of spatial uniformity of the subsample of 23 galaxies brighter than this magnitude still shows considerable skewness; the probability level, due largely to the relative smallness of the sample, rises to 0.025, and so is still significant by conventional standards, although not strongly so.

(d) Because the galaxy apparent magnitudes are model-dependent, in particular the measuring aperture was determined by Peterson in accordance with the expansion hypothesis, completeness out to a prescribed limiting magnitude is likewise model-dependent. Consequently, the Peterson sample is not necessarily complete out to a fixed limiting magnitude under the chronometric hypothesis; the aperture corrections may significantly affect the relative apparent brightness of galaxies near the limiting magnitude. These effects are, however, unlikely to exceed 0.5 mag, and it seems quite safe to suppose that a sample which is complete out to a given limiting magnitude m under the expansion hypothesis is also complete out to a brighter magnitude $m - 1$ under the chronometric hypothesis, with the same metric diameter and a fixed zero-point adjustment of the magnitude scale, so that similar numbers of galaxies are involved in samples complete out to given limits. A limiting magnitude of 13.2 on the chronometric scale was therefore adopted, as comparable to the limiting magnitude of 14 in the expansion hypothesis; this, in fact, selected the identical subsample of 23 galaxies.

The application of the Schmidt V/V_m test within the chronometric framework to this subsample accepts the hypothesis of spatial uniformity, at a highly satisfactory probability level.

(e) The results indicated in (b)–(d) suggest that apparent inhomogeneities in the radial component of the spatial distributions of galaxies may be due to the mode of analysis, and specifically to the employment of the expansion hypothesis, rather than to actual spatial nonuniformity.

The main quantitative results are given in Table 3, whose columns are as follows: (1) is the Abell cluster number. (2) is the measured visual magnitude m at an aperture appropriate to a fixed metric diameter of 20 kpc, on the basis of the Friedman model with $q_0 = \frac{1}{2}$, as reported by Peterson. (3) is the actual semidiameter (radius) of the observed region on the basis of the chronometric hypothesis, with the assumption that

$$H = 100 \text{ kmsec}^{-1} \text{ Mpc}^{-1}$$

at 15 Mpc. (4) is the visual magnitude $m + \Delta m$, corrected on the chronometric hypothesis to a fixed metric diameter of 20 kpc, with the use of the Reynolds–Hubble–Abell surface brightness law earlier indicated. (5) is the mean net aperture correction $\Delta m - \overline{\Delta m}$. (6) is the V/V_m implied by the Hubble model ($m = 5 \log z + \text{const}$, Euclidean space). (7) is the ratio of the number of galaxies in the sample whose V/V_m does not exceed the value in column (7), to the total sample number (spatial uniformity means precisely that (6) – (7) tends to zero as the sample size increases indefinitely). (8) is the actual difference (6) – (7), whose maximum absolute value is the Kolmogorov–Smirnov statistic D. (9), (10), and (11) are the same as (6), (7), and (8), for the subsample of 23 galaxies with the expansion-theoretic apparent magnitude (column (2)) brighter than 14. (12), (13), and (14) are the same within the chronometric hypothesis, for the subsample whose chronometric-theoretic magnitude (column (4)) is brighter than 13.2 (approximately corresponding to the cutoff at magnitude 14 for the expansion-theoretic magnitude, and leading to the identical subsample).

TABLE 3
Analysis of spatial uniformity for Peterson galaxies

1	2	3	4	5	6	7	8	9	10	11	12	13	14
76	13.83	3.84	13.10	0.06	0.20	0.48	−0.28	0.79	0.91	−0.12	0.87	0.89	−0.02
119	14.28	3.57	13.52	0.12	0.37	0.68	−0.31						
147	14.74	3.58	13.98	0.12	0.70	0.95	−0.25						
151	14.29	3.32	13.46	0.19	0.38	0.70	−0.32						
194	12.34	5.45	11.93	−0.23	0.03	0.07	−0.04	0.10	0.13	−0.03	0.16	0.13	0.03
262	12.54	5.59	12.15	−0.25	0.04	0.18	−0.14	0.13	0.17	−0.04	0.22	0.17	0.05
347	12.61	5.45	12.20	−0.23	0.04	0.18	−0.14	0.15	0.26	−0.11	0.24	0.26	−0.02
376	14.72	3.43	13.92	0.16	0.68	0.91	−023						
400	13.21	5.01	12.73	−0.16	0.09	0.34	−0.25	0.34	0.57	−0.23	0.51	0.57	−0.06
407	14.70	3.47	13.91	0.15	0.66	0.89	−0.23						
426	12.13	5.38	11.71	−0.22	0.02	0.05	−0.03	0.08	0.09	−0.01	0.12	0.09	0.03
505	14.43	3.27	13.59	0.20	0.46	0.82	−0.36						
539	13.74	4.41	13.16	−0.06	0.23	0.52	−0.29	0.70	0.87	−0.17	0.95	0.96	−0.01
548	14.13	3.77	13.42	0.07	0.30	0.59	−0.29						
569	12.71	5.27	12.27	−0.20	0.04	0.18	−0.14	0.17	0.35	−0.18	0.27	0.30	−0.03
576	14.37	3.72	13.65	0.06	0.42	0.77	−0.35						
634	13.61	4.50	13.05	−0.08	0.15	0.41	−0.26	0.58	0.78	−0.20	0.81	0.78	0.03
671	14.23	3.40	13.42	0.17	0.35	0.66	−0.31						
754	14.34	3.29	13.50	0.20	0.40	0.75	−0.35						
779	13.22	5.13	12.76	−0.18	0.09	0.34	−0.25	0.34	0.61	−0.27	0.53	0.61	−0.08
993	14.48	3.30	13.65	0.19	0.49	0.84	−0.35						
1060	11.60	6.70	11.34	−0.38	0.01	0.02	−0.01	0.04	0.04	0	0.07	0.04	0.03
1139	14.20	3.84	13.50	0.06	0.33	0.61	−0.28						
1185	13.58	3.97	12.91	0.03	0.14	0.39	−0.25	0.56	0.74	−0.18	0.67	0.70	−0.03
1213	14.24	4.34	13.65	0.05	0.35	0.66	−0.31						
1228	14.10	4.00	13.44	0.02	0.29	0.57	−0.28						
1257	14.52	4.02	13.86	0.02	0.52	0.86	−0.34						
1314	13.65	4.04	13.00	0.01	0.16	0.43	−0.27	0.62	0.83	−0.21	0.76	0.74	0.02
1318	12.75	5.28	12.31	−0.20	0.05	0.20	−0.15	0.18	0.39	−0.21	0.28	0.35	−0.07
1367	12.56	5.09	12.10	−0.18	0.04	0.18	−0.14	0.14	0.22	−0.18	0.87	0.89	0
1377	14.73	3.35	13.90	0.18	0.69	0.93	−0.24						
1656	12.69	4.84	12.19	−0.14	0.04	0.18	−0.14	0.16	0.30	−0.14	0.24	0.26	−0.02
1736	14.33	3.62	13.58	0.11	0.40	0.75	−0.35						
2052	13.83	3.96	13.16	0.03	0.20	0.48	−0.28	0.79	0.96	−0.17	0.95	0.96	−0.01
2147	13.86	3.96	13.19	0.03	0.21	0.50	−0.29	0.82	1.00	−0.18	0.99	1.00	−0.01
2151	14.04	3.97	13.37	0.03	0.27	0.55	−0.28						
2152	14.40	3.58	13.64	0.12	0.44	0.80	−0.26						
2162	13.48	4.15	12.85	−0.01	0.12	0.36	−0.24	0.49	0.70	−0.21	0.62	0.65	−0.03
2197	13.20	4.12	12.56	0	0.09	0.34	−0.25	0.33	0.52	−0.19	0.40	0.48	−0.08
2199	13.02	4.14	12.39	−0.01	0.07	0.25	−0.18	0.26	0.48	−0.22	0.31	0.43	−0.12
2319	14.78	3.25	13.93	0.21	0.74	1.00	−0.26						
2634	13.22	4.18	12.60	−0.02	0.09	0.34	−0.25	0.34	0.65	−0.31	0.43	0.52	−0.09
2657	14.77	3.64	14.03	0.10	0.73	0.98	−0.25						
2666	12.95	4.45	12.36	−0.05	0.06	0.23	−0.17	0.23	0.43	−0.20	0.30	0.39	−0.09

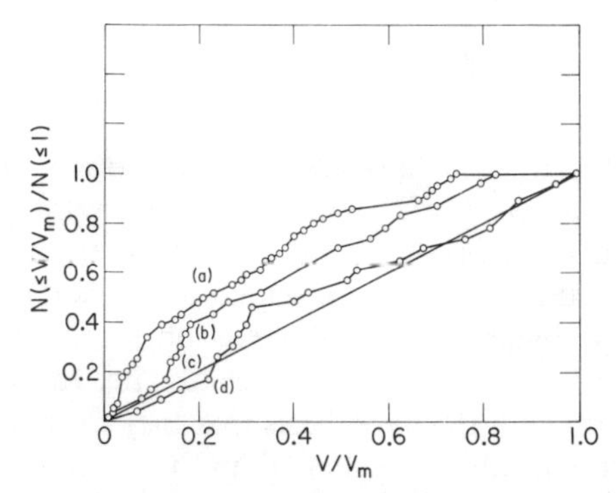

Figure 5 *The V/V_m test for the Peterson sample.*
○, individual galaxies, values computed on the following bases: (a) Hubble theory, limiting magnitude 15 (all 44 galaxies); (b) Hubble theory, limiting magnitude 14 (23 galaxies) (d) chronometric theory, limiting magnitude 14.2 (same 23 galaxies as in (b)). (c) Theoretical line for radial spatial uniformity. Thus apparent radial spatial uniformity is materially a function of the theory employed.

The only part of the aperture correction that is relevant to cosmological testing is the deviation from the mean correction given in column (5); this is small, having a root mean square of 0.15 mag. The respective Kolmogorov–Smirnov statistics D for the three cases (i.e., the maxima of the absolute values of the entries in columns (8), (11), and (14), respectively) are 0.36, 0.31, and 0.12. Assuming a spatially uniform population, the respective probabilities P of obtaining values of D this large are 2×10^{-5}, 0.024, and >0.5, employing here the asymptotic law $P \sim 2\exp(-2nD^2)$, where n is the sample size. This formula is asymptotic as $n \to \infty$; however, it is considered to give a good approximation for relatively small values of n; the results are consistent with the confidence intervals given in Pearson and Hartley (1972).

Some of the results are summarized in Figure 5, in which the fraction observed having V/V_m less than a given value λ is compared with λ, in each of the three statistical situations under consideration here. The abscissa is then V/V_m, or the volume out to the redshift of the object is the theory in question, divided by the maximum volume within which the object would remain in the sample, according to the theory. The ordinate is the cumulative frequency, expressed as a fraction of the total number of objects in the sample, of objects whose V/V_m is not exceeded by the abscissa. The straight line segment between (0, 0) and (1, 1) is the theoretical line expected for an infinitely large sample of objects uniformly distributed in space; above this line, in order, come the observed line for the subsample of 23 galaxies limiting magnitude 13.2 according to the chronometric theory; the subsample of limiting magnitude 14 according to the Hubble theory (actually the same as just indicated), and the entire sample (44 galaxies) according to the Hubble theory.

7. Markarian galaxies and N-galaxies

A relatively objectively defined sample of galaxies of the former type has been observed by Sargent (1972). These galaxies resemble Seyfert galaxies, and aperture effects for them should be relatively marginal. It is therefore of interest to compare the (m, z) pairs observed by Sargent with the theoretical m–z relation for a single type of luminous object, for which the present galaxies appear to be a relatively good candidate. In any event, it seems appropriate to begin with the hypothesis that they form a single luminosity class as the simplest tenable one. It is found that the chronometric curve, adjusted to the average absolute magnitude of the galaxies as given by the chronometric theory, fits within a dispersion of 0.77 mag the Sargent observations. The corresponding dispersion as given by the Hubble theory is 1.12; the sample dispersion is 0.87. The qualitative point here is not so much that the chronometric dispersion is less than the expansion-theoretic dispersion; the sample is too small for statistical significance, although in conjunction with other samples presented here it is statistically relevant. Rather it is the surprising excess of the expansion-theoretic over the sample dispersion; an excess of the magnitude here found seems quite unlikely, on the expansion-theoretic hypothesis, but to make a formal statistical analysis would seem supererogatory, in view of the weight of other evidence and the always possible defense of unknown selection effects. Not only does the chronometric theory supply a reasonable model, and one distinctly better than the expansionary one in this instance, but it also alters in a reassuring way the absolute luminosity of these objects vis-à-vis quasars, as well as vis-à-vis the classical Seyferts. In many important respects, other than their absolute luminosities as given by the expansionary theory, these objects appear very closely related, and possibly substantially identical. However, on the expansion theory, quasars are several magnitudes brighter than the present Markarian galaxies, and these in turn are according to Sargent (1971) brighter than the classical Seyfert galaxies. These apparent differences in luminosity are seen to result from the theoretical analysis, and to be not necessarily real, by an analysis from the chronometric standpoint, according to which there is little difference between the intrinsic luminosities of these three groups of objects. See also Rees and Sargent (1972).

Figure 6 exemplifies some of these points. The DeVeny quasar lines are based on the sample of 158 quasars described later; the magnitude–redshift theoretical curves for these have been corrected (in accordance with an oral communication from W. L. W. Sargent) to represent photographic magnitudes comparable to those reported in Sargent (1972) by taking $m_p \sim m_v + 0.4$ for the present galaxies.

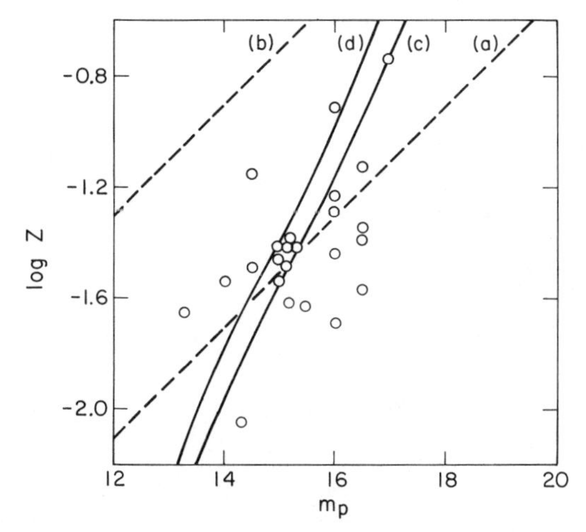

Figure 6 *The redshift–magnitude relation for Seyfert-like Markarian galaxies studied by Sargent* (1972).

(a) Best-fitting Hubble line to present galaxies ($\sigma = 1.12$); (b) best-fitting Hubble line to quasars studied by DeVeny *et al.* (1971); (c) best-fitting chronometric curve to present galaxies ($\sigma = 0.77$); (d) best-fitting chronometric curve to DeVeny quasars. In particular, quasars and Seyfert-like galaxies have little difference in intrinsic luminosity on the chronometric hypothesis, although the difference is quite large on the expansion hypothesis.

An analogous situation is presented by the N-galaxies. Their close relationship to quasars has been remarked by many authors, and Lynden-Bell (1971) has proposed that they constitute "miniquasars," similar to but less luminous than quasars. On the expansion hypothesis, N-galaxies average ~ 3 magnitudes fainter than the average quasar (as represented by the DeVeny list), but in other important respects they resemble quasars. The fact is that on the chronometric hypothesis, the N-galaxies have average intrinsic luminosity within 0.5 mag of the average quasar. Moreover, the chronometric m–z curve fits the N-galaxy data with a distinctly smaller dispersion than does the Hubble line.

Admittedly, the number of N-galaxies having reliable magnitudes and redshifts is too small for the difference in dispersion to be statistically significant, but in conjunction with the other considerations of this section, the data for N-galaxies provide a measure of support for the chronometric hypothesis. Figure 7 shows the redshifts and magnitudes for the N-galaxies considered by Sandage (1967). The corrected magnitudes given by Sandage have been used, and the optically highly variable galaxy 3C 391, as reported by Sandage (1967), has been excluded. The standard deviations of the residuals of the observed magnitudes from the best-fitting theoretical lines are 0.49 for the chronometric theory and 0.68 for the expansion theory.

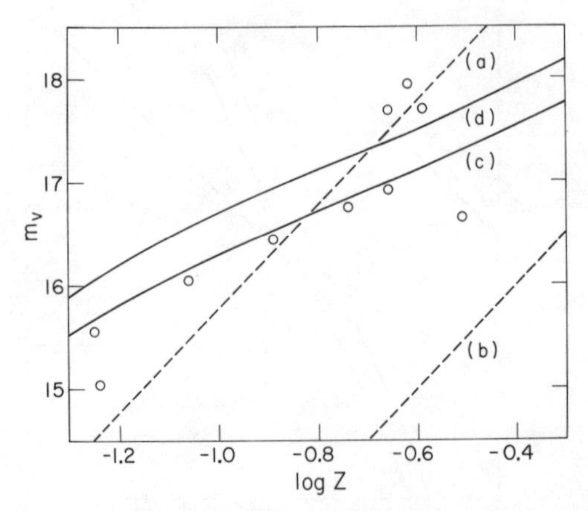

Figure 7 *The redshift–magnitude relation for* N-*galaxies listed by Sandage* (1967). (a) Best-fitting Hubble line to present galaxies ($\sigma = 0.68$); (b) best-fitting Hubble line to quasars studied by DeVeny *et al.* (1971); (c) best-fitting chronometric curve to present galaxies ($\sigma = 0.51$); (d) best-fitting chronometric curve to DeVeny quasars. Again, these galaxies differ little from quasars in intrinsic luminosity on the chronometric hypothesis, but differ substantially on the expansion hypothesis.

The interpretation of these results within the chronometric theory, as regards the relation between quasars on the one hand and Seyfert-like or N-galaxies on the other is necessarily rather speculative, and of a different nature from the considerations involved in systematic hypothesis testing. Nevertheless it may not be amiss and indeed is probably peripherally relevant to note the indication that such galaxies are not only similar to but are perhaps identical with a certain category of quasar of average intrinsic luminosity. That is, if at larger redshifts, many of these galaxies might well appear to be quasars. Observations on spectral functions required for material confirmation may not be available for some time, but it may be noted parenthetically that the analysis of quasar observations (cf. below) provides some circumstantial evidence: (a) chronometrically there is a statistically insignificant but nevertheless noticeable deficiency of quasars in the redshift range 0.0–0.3, which could be removed by hypothesizing the identity of N- and certain Seyfert-like galaxies with certain classes of quasars; (b) the model-independent distribution of luminosities of quasars (cf. below) does not deviate in a statistically significant way from a normal distribution, but there are nevertheless some clearly marked groupings suggesting that it is more precisely a superposition of normal distributions, of effectively non-overlapping ranges. The brightest fifth of the quasars in the DeVeny list ("brightest" in a model-independent sense detailed below) have an optical luminosity ~ 1.2 mag brighter than the average quasar in the list; and there is a noticeable gap in luminosity between these bright quasars and the average ones. The latter quasars thereby appear to constitute "miniquasars" relative to the bright ones, in a sense analogous to that employed by Lynden-Bell (1971), whose theoretical proposal, expanded and modified in the fashion just suggested, appears to be in agreement with present observations. The recent observations of Sandage (1973) on N-galaxies lend further support to the conjecture that N-galaxies at higher redshifts may appear as quasars.

That these results are not reflections of small sample size or of coincidental selections is confirmed by the study of a substantial sample of Markarian galaxies listed by Arakelyan *et al.* (1972). These are largely at the higher redshifts thought to be beyond the local supercluster postulated by some in order to reconcile the apparent square-law dependence of redshift on distance for low-redshift galaxies with the expanding-universe theory. No special selection effects relatively favorable to a square law are known for these data. However, the square law decreases the dispersion in apparent magnitude, while the linear law increases it. Specifically, the dispersions in apparent magnitude (a) and in absolute magnitude based respectively on the (b) chronometric prediction (differing trivially from the square law in this redshift range) and (c) Hubble law, are as follows. For the 60 galaxies with $cz > 3000$ km sec^{-1} (average value, 8000), (a) 0.89, (b) 0.84, (c) 1.04. For the full sample of 69 galaxies, (a) 1.09, (b) 1.01, (c) 1.51. The results are qualitatively unaltered if the galaxies are arranged in order of increasing redshift, and divided into bins containing equal numbers of galaxies, the brightest, second brightest, etc., in each bin being selected; or if the observations at extreme redshifts are deleted from the sample.

8. The redshift–magnitude relation for nearby galaxies

The major study by G. de Vaucouleurs (1972) of the redshift–magnitude relation for about 100 nearby groups of galaxies confirms the apparent quadratic dependence of redshift on distance, which was noted by Hawkins (1962) on the basis of the observations of Humason *et al.* (1956) (regarding historical origins, cf. also Lundmark, 1920, 1925). This is consistent with the chronometric theory, but deviates from the law of Hubble (1929).

It has been proposed by de Vaucouleurs that the expansion theory is basically correct, but that a local spatial anisotropy distorts the redshift–magnitude relation. The hierarchical model proposed by de Vaucouleurs is related in direction to that originally proposed by Charlier, as well as more recent ideas of Holmberg; it appears to be in satisfactory agreement with low-redshift observations for the m–z relation of galaxies. It is, however, scientifically less economical that the chronometric theory, in that the latter involves no sacrifice of spatial homogeneity or additional parameters.

As discussed by G. de Vaucouleurs (1972), there is a persistent anomaly in the determination of Hubble's parameter by different observations, and specifically between the lower values obtained from observations of Virgo cluster objects and the higher values obtained from observations including the present redshift–magnitude data. A further advantage of the chronometric over the expansion-theoretic model is that it reconciles the different values on the basis of the different distances to the objects under observation. Thus the velocity/distance relation of these groups is apparently nonlinear for $\Delta < 30$ Mpc. The velocity/distance ratio increases from $H \cong 50$ to 150 km sec^{-1} Mpc^{-1} when Δ increases from $\Delta \cong 5$ to $\Delta \cong 25$ Mpc, according to de Vaucouleurs (1972). On the chronometric hypothesis, the value $H = 50$ at a distance $\Delta = 5$ is equivalent to the value $H = 86$ at $\Delta = 15$; similarly the value $H = 150$ is equivalent to the value $H = 116$ at $\Delta = 15$; they are thus within 16% of the value $H = 100$ at $\Delta = 15$ Mpc which has been adopted in the present work. This is a level of accuracy comparable with optimistic informed estimates of the attainable accuracy (cf. Sandage, 1970). It is relevant to note also that one of the most scrupulous estimates of the Hubble parameter, that due to Holmberg (1964), of 80 km sec^{-1} Mpc^{-1}, while not specifically based on a particular value for the distance Δ, may reasonably be considered to correspond to $\Delta \sim 10$; it is then equivalent to the value $H = 98$ at $\Delta = 15$. This differs insignificantly from the value $H = 100$ at $\Delta = 15$ employed here, as does the eclectically based estimate $H = 95$ by van den Bergh (1970). From the chronometric standpoint, there is thus no significant anomaly in the differing values of the Hubble parameter as determined by most leading investigators. The only exception, the recent determination $H \sim 50$ by Sandage, is based on quite different observations and new distance scales, and seems explicable on this basis. The work of Abell (1972) emphasized primarily the uncertainty in the Hubble parameter; a possible low value for H is cited basically as an illustration of the dependence of its determination on the assumption made regarding the comparative luminosity function of the Virgo cluster; and the difficulty of resolution of the fundamental question of an operational and model-independent selection procedure for "cluster" tends seriously to moot statistically cosmological cluster samples.

The foregoing indications regarding the phenomenology of low-redshift galaxies suggest a comprehensive statistical analysis of the de Vaucouleurs tape, representing an updating to 1972 of the material in G. de Vaucouleurs and A. de Vaucouleurs (1964). Included here are many more galaxies than those on which the Hubble law was originally based. Such an analysis was conducted jointly with J. F. Nicoll, employing all of the data of objectively delineated subsamples, entirely without corrections or other uncertain emendations, and standard contemporary principles of statistical estimation and hypothesis testing. The results are extremely favorable to the square redshift–distance law, both at the model-building and hypothesis-testing levels. They are quite unfavorable to the Hubble law at the model-building level, but at the hypothesis testing level the law may be marginally acceptable, with some emendations.

Making the purely phenomenological assumption that $z \propto r^p$, where z is the redshift and r the distance, apart from peculiar motions, for some constant exponent p, and sufficiently small distances r, simply embodies the observed facts that redshifts generally increase with distance and vanish near the Galaxy. Statistically, it is assumed that bright galaxies form a true statistical population, at least for redshifts $\leqq 0.03$, i.e., there is no evolution in this range. Where relevant it is assumed further that the spatial distribution of the galaxies is radially homogeneous; no assumption as to isotropy is required. It then follows that the probability density for r varies as $r^2\,dr$, whence that for z varies as $z^{3q-1}\,dz$, where $q = p^{-1}$.

The exponent p may be estimated from observed relations between the magnitudes, redshifts, and angular diameters of galaxies, in accordance with the maximum-likelihood procedure. For any value of p, the apparent luminosity will vary with the absolute luminosity in accordance with the inverse square law, the redshift factor $(1 + z)^{-1}$, and possible theory-dependent factors which may be presumed negligible for $z \leqq 0.03$, as they surely are for all theories considered realistic. The apparent angular diameter θ will similarly vary with the absolute diameter Δ, and inversely with r. It follows that $m = 5q \log z + M$ and $\log \theta = -q \log z + A$, where $A = \log \Delta$, apart from terms of order z or less, which are here negligible. Absolute magnitudes M and logarithmic diameters A may be defined by these equations.

The joint probability distribution of z, M, and A takes the form $P(z)P(M, A)$, in view of the stochastic independence of M and A from z, where $P(z) = Cz^{3q-1}$ with $C = 3q[z_2^{3q} - z_1^{3q}]^{-1}$ if the redshift interval under consideration is $z_1 < z < z_2$, and $P(M, A)$ takes the form in terms of observed quantities: $P(m - 5q \log z - \bar{M}, \log \theta + q \log z - \bar{A})$, where $\bar{M}$ and $\bar{A}$ are the population means. The unknown function $P(M, A)$ will be assumed otherwise to depend only on the standard deviations σ_M and σ_A of M and A, and their correlation ρ, by the normal law. This is standard phenomenological procedure; serves to ensure the coincidence of maximum-likelihood and least-square estimation for the parameter involved in $P(M, A)$; and may be confirmed by statistical testing of the resulting sample distributions of absolute quantities.

The various bi- and univariate distributions then follow by integration, and depend on parameters that are functions of the foregoing. The maximum-likelihood procedure consists in choosing the parameters to maximize the corresponding probability density for the observed sample. With an imposed cutoff in apparent magnitude of m_L, the new probability density $P(z, M, A)$ is derived by multiplication of $P(z)P(M, A)$ by $(\iiint_{m \leq m_L} P(z)P(M, A)\, dz\, dM\, dA)^{-1}$ for $m \leqq m_L$, and by 0 for $m > m_L$. The corresponding maximum-likelihood estimates cannot be given in analytically explicit form, but are determinable by successive approximation procedures.

In addition to the familiar m–z, θ–z, and $N(z)$ relations, it is interesting to consider the $N(V/V_{\mathrm{m}})$ relation. The original Schmidt V/V_{m} test (see Schmidt, 1968) involved no a priori limitations on the redshifts involved, but it is essential for observational reasons (incompleteness in redshift determinations for larger z), as well as to enhance its discriminatory capacity, to adapt it to the case in which it is a priori required that $z_1 < z < z_2$, where z_1 and z_2 are given. The V $(= V(z))$ then excludes the region up to the redshift z_1, so $V(z) \propto z^{3q} - z_1^{3q}$, while the V_{m} is $V(z_{\mathrm{m}})$, where z_{m} is given by the equation $m_L - m = (5q)\log(z_{\mathrm{m}}/z)$ provided z_{m} is determined from this equation is $\leqq z_2$; otherwise $z_{\mathrm{m}} = z_2$ and $V_{\mathrm{m}} = V(z_2) - V(z_1)$. Thus

$$V/V_{\mathrm{m}} = (z^{3q} - z_1^{3q})(z_{\mathrm{m}}^{3q} - z_1^{3q})^{-1},$$

where z_{m} is the indicated p-dependent function of m, z, m_L, and z_2. The principle of the Schmidt test, i.e., the uniform distribution in $[0, 1]$ of V/V_{m}, on the assumption of radial spatial homogeneity, applies equally well to this generalized situation; and, unlike the original case ($z_1 = 0$, $z_2 = \infty$), the test is in practice effectively discriminatory between different redshift–distance relations, even when applied to low-redshift objects.

The deviation of an observed from the theoretical uniform distribution can be measured by the Kolmogorov–Smirnov statistic D, which is the maximum absolute difference between the cumulative observed and theoretical distributions. Alternatively, an approximately normal statistic X similar to that employed by Schmidt (1968) is given by the mean of the V/V_{m}, centered to zero mean and normalized to unit variance, i.e.,

$$X = (12/N)^{1/2} \sum (V/V_{\mathrm{m}} - \tfrac{1}{2});$$

however, X may vanish although the distributions are different. In either case, the value of p that minimizes the deviation (and so maximizes the corresponding probability) provides an analogue to the maximum-likelihood estimate, and the corresponding confidence intervals for p effectively substitute for dispersions in the estimates.

The de Vaucouleurs catalog includes galaxies having redshifts up to $\sim cz = 10{,}000$ and magnitudes up to ~ 15, and is estimated to be overall $\sim 50\%$ complete out to a magnitude of 13. Completeness is probably much greater in limited redshift regions, and its approximate validity out to $cz \sim 2000$ is suggested by the observed $N(< z)$ relation shown in Figure 8.

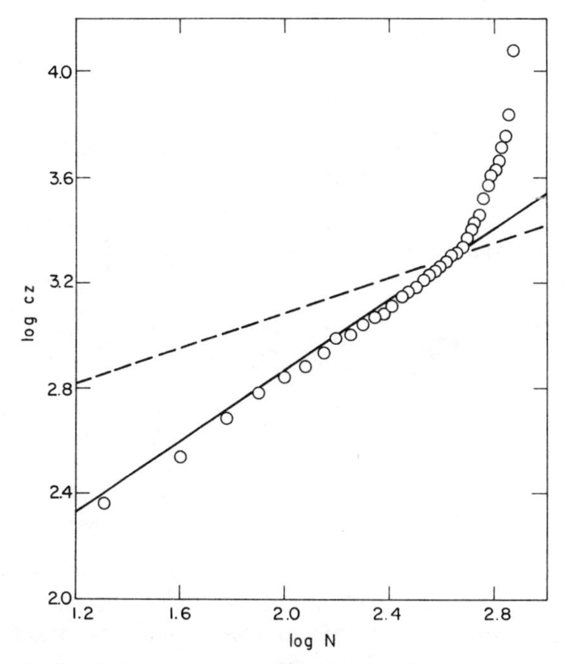

Figure 8 *The* log N–log z *relation for all galaxies included in the de Vaucouleurs tape, having m–z–θ data* (742 *galaxies*).

○, observational points; — and ---, the lines $\log N\ (\leqq cz) - \log N\ (\leqq 2000) = (3/p)\log(cz/2000)$, for the values $p = 2$ and $p = 1$, respectively. These represent the theoretical intrinsic number of galaxies in the indicated redshift regions, on the assumption of a uniform spatial distribution of galaxies. Progressive incompleteness in redshift determinations is anticipated for $z \gg 0$, and is indicated by the strong deviation in this range of the observational relation from the very nearly linear one found for $cz \lesssim 2000$.

For greater conservatism, the test has also been carried out with a limiting magnitude of 12.5. It seems likely that the sample is nearly complete to this limit; indeed results with brighter limits are similar but less definitive because of smaller sample size. For statistical validity, actual completeness is not required, but only randomness within the complete population, in a radial direction, out to the distance corresponding to $cz = 2000$. There is no special reason to doubt that this holds.

a. ***Estimates of*** *p*

In addition to the m–z and m–θ relations, the m–z–θ relation for all 742 galaxies having these data in the de Vaucouleurs catalog were employed for maximum-likelihood estimation. The results regarding ρ are shown in Table 4.

TABLE 4
Maximum-likelihood estimates of the redshift–distance exponent

Sample	Full catalog (742 galaxies)		Subsample with $500 \leqq cz \leqq 2000$, $B(0) \leqq 13$ (350 galaxies)	
Relation	Exponent	Dispersion in exponent	Exponent	Dispersion in exponent
m–z	2.05	0.07	2.39	0.28
θ–z	1.96	0.08	1.92	0.26
m–θ–z	2.04	0.07	2.39	0.29
N–z			2.57	0.31
m–N–z			2.48	0.21
θ–N–z			2.32	0.22
m–θ–N–z			2.48	0.21

In order to assay the sensitivity of the results to conceivable selection effects, estimates were also made for a number of subsamples, selected on redshift range, apparent magnitude, morphological type, and field of observation. No evidence for significant sensitivity was found (cf. the discussion below). A subsample which is representative and reasonable on a priori grounds as well as on the basis of internal indications, is that defined by the limits $500 \leqq cz \leqq 2000$ and $B(0) \leqq 13$. The higher cutoff in cz eliminates a region in which there is a clear phenomenological break in the N–z relation, as shown in Figure 8, and anticipated as a result of incompleteness in redshift determinations for higher redshifts. The results regarding p for this subsample of 350 galaxies, including those based on relations involving N, which would be inappropriate for the full catalog, are shown on the right in Table 4.

The median value of p is 2.39 (the mean is 2.36); the difference from the value $p = 2$ is not statistically significant, in view of the median dispersion of 0.26 (mean of 0.25). However, the excess over 2, as compared with the results from the full sample, is in the direction of a magnitude truncation effect, and indeed the explicit incorporation of an a priori magnitude cutoff into the maximum-likelihood procedure leads to estimates closer to $p = 2$. The relatively lengthy computations for the modified procedure have been carried out for several cutoffs and two redshift intervals on the basis of the observed m–z relation, and are given in Table 5.

In view of the indication from the original maximum-likelihood estimates (in particular, the z-independence of the distribution of residuals from the corresponding theoretical m–z law; cf. Figure 1) that $m = 13$ should be

TABLE 5
Estimates incorporating an a priori magnitude cutoff

Redshift range:	500 ≦ cz ≦ 2000			500 ≦ cz ≦ 1800		
Limiting magnitude	Sample size	p	σ_p	Sample size	p	σ_p
13.00	350	1.86	0.18	303	1.86	0.20
12.90	339	1.86	0.19	293	1.92	0.22
12.85	330	1.95	0.22	288	1.96	0.24
12.80	325	1.94	0.23	286	1.90	0.25

beyond the faintness necessary in the range $cz \leqq 2000$ to include all but a small fraction of the relevant population, the marginal effect of allowance for the magnitude cutoff on the estimates was to be expected.

b. Estimates of galaxy parameters

The maximum-likelihood estimates of the basic parameters of the joint absolute magnitude–diameter distribution are shown in Table 6. For comparison purposes, the same parameters as estimated from the data on the basis of the prior hypotheses that $p = 1$ or 2 are also shown.

For each sample, the estimated parameters are rather insensitive to the relation employed, particularly in the case of the dispersions, which do not

TABLE 6
Maximum-likelihood estimates of galaxy parameters

Sample:	Full catalog (742 galaxies)					Subsample (350 galaxies)				
Relation	$\bar{M}$	σ_M	$\bar{A}$	σ_A	ρ	$\bar{M}$	σ_M	$\bar{A}$	σ_A	ρ
m–z	17.92	0.93	—	—	—	16.79	0.72	—	—	—
θ–z	—	—	0.14	0.23	—	—	—	0.16	0.21	—
m–θ–z	17.94	0.93	0.18	0.23	0.76	16.78	0.72	0.41	0.21	0.70
m–N–z	—	—	—	—	—	16.60	0.72	—	—	—
θ–N–z	—	—	—	—	—	—	—	0.38	0.21	—
m–θ–N–z	—	—	—	—	—	16.60	0.72	0.44	0.21	0.70
$p = 2$[a]	18.05	0.93	0.16	0.23	0.76	17.77	0.72	0.21	0.21	0.70
$p = 1$[a]	23.72	1.36	−0.97	0.30	0.87	23.76	0.85	−0.99	0.22	0.74

[a] Here a value of p is assumed a priori, and the maximum-likelihood procedure is applied to the other parameters (least-square estimation in these cases).

differ within the accuracy quoted between their values for $p = 2$ and the maximum-likelihood estimates of p. As anticipated from the cutoff on apparent magnitude in the subsample, it has generally smaller dispersions and brighter mean magnitudes than the full catalog. For $p = 1$, the dispersions are distinctly larger than for $p = 2$, and are in fact generally larger than corresponding ones in the raw data. Specifically, the latter are, for the full sample, $\sigma_m = 1.33$ and $\sigma_{\log\theta} = 0.30$; for the subsample, $\sigma_m = 0.79$ and $\sigma_{\log\theta} = 0.22$. This negative predictive power for the $p = 1$ assumption is equally the case for the N–z relation for the subsample, where $D = 0.27$ for $p = 1$ and $D = 0.04$ for the deviation of the observed relation from a law of uniform distribution in redshift.

The allowance for the observational cutoff in magnitude has naturally the effect of increasing the estimated dispersion in absolute magnitude as well as the estimated mean magnitude. The results of this more refined analysis are $\sigma \sim 0.92$ and $\bar{M} \sim 18.1$, for all of the samples in Table 6. It is interesting that these values do not differ significantly from those estimated for the full sample without allowance for the observational magnitude truncation.

c. *The Schmidt V/V_m test*

As earlier indicated, this test is independent of magnitude truncation, whether in all of space or in a fixed redshift interval, as is here appropriate. The influence of peculiar velocities may be largely suppressed by elimination of sufficiently low redshifts. In view of the earlier-cited dispersion of < 100 km sec^{-1} among blueshifted galaxies in the catalog, the elimination of galaxies with redshifts $\leqq 500$ km sec^{-1} appears likely to achieve this end. At the same time it should serve to avoid ultralocal irregularities.

At the other extreme, the dependency on redshift of incompleteness in redshift determinations requires the elimination of correspondingly high redshifts. The close approximation to linearity of the phenomenological $\log N(< z)$–$\log z$ relation up to but not beyond the limit $cz = 2000$ km sec^{-1} (cf. Figure 8), provides objective indications for the appropriateness of this redshift as an upper limit. Consequently the redshift limits $500 \leqq cz \leqq 2000$ have been adopted in the tests detailed here. Computations for slightly different ranges bounded by $cz = 300$ at the lower range and $cz = 1800$ at the upper, have shown insensitivity to the precise limits employed.

As earlier indicated, it is problematical whether the catalog galaxies brighter than 13^m in the redshift range $\leqq 2000$ form a random subsample of all such galaxies. However, this seems likely to be effectively the case with a limiting magnitude of 12.5. The results for both limiting magnitudes are given in Table 7.

TABLE 7
Maximum-probability estimates of the redshift–distance exponent from the $N - V/V_m$ relation

Redshift interval	Limiting magnitude	D estimate	X estimate
500–2000	13.0 (350 galaxies)	2.24	2.25
	12.5 (286 galaxies)	2.17	2.05
500–1800	13.0 (303 galaxies)	2.20	2.21
	12.5 (254 galaxies)	2.01	1.92
300–2000	13.0 (379 galaxies)	2.08	2.09
	12.5 (312 galaxies)	1.82	1.84
300–1800	13.0 (332 galaxies)	2.05	2.05
	12.5 (280 galaxies)	1.75	1.75
0–2000	13.0 (409 galaxies)	1.93	2.00
	12.5 (340 galaxies)	1.77	1.81

Confidence intervals for these estimates may be determined, and the V/V_m procedure clarified, by reference to Table 8, which gives for each limiting magnitude and a range of values of p the corresponding values of D, the probability $P(D)$ of obtaining a deviation as large as D, and X. In particular, with the more conservative limiting magnitude of 12.5, the hypothesis that $p = 1$ leads to probabilities of deviations as large as those

TABLE 8
Deviations from spatial uniformity as indicated by the $N - V/V_m$ relation for galaxies with $500 \leqq cz \leqq 2000$

	Limiting magnitude 13			Limiting magnitude 12.5		
p	D	$P(D)$	X	D	$P(D)$	X
10.0	0.165	0.00000001	6.15	0.134	0.00007	4.27
5.0	0.114	0.0002	4.09	0.099	0.008	2.95
3.0	0.064	0.110	1.83	0.057	0.304	1.52
2.5	0.042	0.570	0.72	0.041	0.726	0.84
2.3	0.029	0.928	0.16	0.032	0.924	0.50
2.2	0.029	0.931	−0.15	0.026	0.989	0.31
2.1	0.030	0.904	−0.47	0.025	0.995	0.11
2.0	0.038	0.684	−0.83	0.030	0.960	−0.11
1.8	0.049	0.363	−1.58	0.041	0.734	−0.58
1.6	0.075	0.036	−2.43	0.058	0.296	−1.12
1.4	0.091	0.006	−3.43	0.071	0.115	−1.70
1.2	0.119	0.0001	−4.60	0.078	0.061	−2.37
1.0	0.149	0.0000004	−5.92	0.104	0.004	−3.11

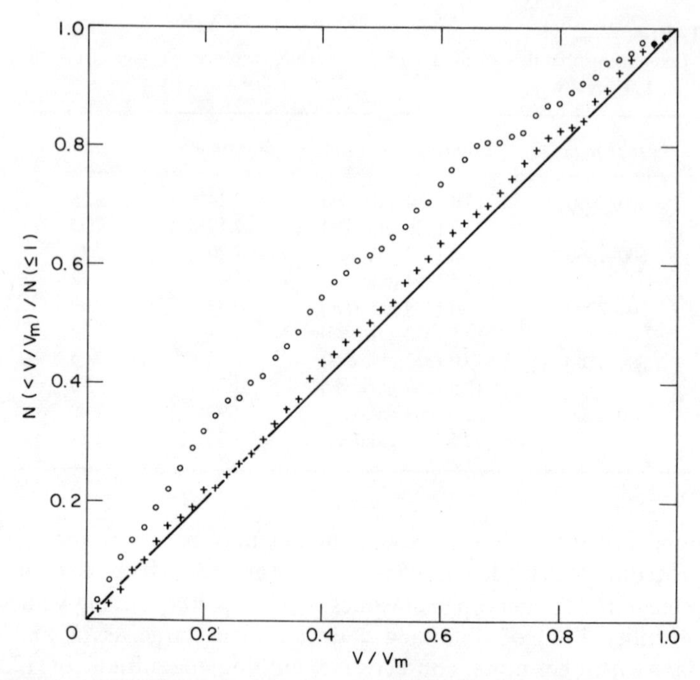

Figure 9 *The $N(< V/V_m)$ relation for the subsample of the galaxies included in the de Vaucouleurs tape in the redshift range* $500 \leqq cz \leqq 2000$, *and not fainter than* 13^m, *on the basis of the* $p = 1$ *and 2 hypotheses* (○ *and* ×, *respectively*).

The deviation from spatial uniformity (—) for a hypothetical linear redshift–distance law is highly significant for a sample of the present size (350 galaxies); in the case of a hypothetical square law, the agreement is quite satisfactory.

observed in D and X of 0.004 and 0.001, respectively; for $p = 2$, the corresponding probabilities are 0.96 and 0.83. Figure 9 shows the sample $N(V/V_m)$ relation for $p = 1$ (open circles) and $p = 2$ (crosses) together with the line representing theoretical spatial uniformily in the radial direction.

d. Discussion

On a straightforward phenomenological basis, the results strongly support a value of $p \sim 2$, and reject the value $p = 1$. As in virtually any situation in which controlled random sampling is inherently difficult, some refinements in procedure might be contemplated. The major ones, and the only ones that appear to have nontrivial potential for alteration of the main conclusion are: (a) selection on morphological type; (b) limitation of the region of the sky in order to reduce the possible effects of different telescope locations and parameters.

Comprehensive quantitative examination of the possible effect (a) shows that it is not real. The relative strength of the indications for the value $p = 2$ as against the value $p = 1$ is quite unaffected by selection on morphological type. For the most refined estimates, i.e., those that allow explicitly for an a priori magnitude cutoff, a substantial sample is necessary to ensure proper convergence of the successive approximations procedure; consequently, the subsample of galaxies that are either elliptical or spiral was studied in this connection. The results, summarized in Table 9, which is comparable to Table 5 treating all types of galaxies, yield an average estimate of $p = 1.88 \pm 0.21$, based as earlier on the subsample in the restricted redshift range $500 \leqq cz \leqq 2000$.

TABLE 9
Maximum-likelihood estimates incorporating an a priori magnitude cutoff, for the subsample of all galaxies with data which are either elliptical or spiral, and have $500 \leqq cz \leqq 2000$

Limiting magnitude	Sample size	Estimate of p	Dispersion in p
13.00	271	1.89	0.20
12.95	268	1.87	0.20
12.90	264	1.85	0.20
12.85	259	1.89	0.22
12.80	255	1.89	0.24

Results for morphologically selected subsamples of the full sample having m–z–θ data, are generally quite similar. The results are summarized in Table 10, which is comparable to Table 4, some of whose results are repeated for ready comparison. For the classic m–z relation, which in the present analysis appears the most stable, the estimate of p, averaged with equal weight over the subsamples of ellipticals, spirals, and lenticulars, is 2.03 ± 0.26. No less compelling is the scrutiny of the resulting dispersions. Those from the $p = 1$ law are larger than those in the apparent quantity observed, whether magnitudes or logarithmic diameters, for most of the samples, and appear never to be significantly less than those in the raw data. On the other hand, the dispersions from the $p = 2$ law are quite materially less than those in the apparent quantities, and do not differ appreciably if at all from the dispersions from the optimal value of p.

Similarly, potential effect (b) is not quantitatively visible on separate analysis of the northern and southern hemispheres, galactic or celestial. The results are summarized in Table 11, which is statistically comparable to Tables 4 and 10, all galaxies with m–z–θ data in the indicated portion of the

TABLE 10
Effect of selection on morphological type

Type	Estimates of p			Dispersions in magnitude				Dispersions in log A			
	m–z	θ–z	m–θ–z	σ_m	σ_{M_1}	σ_{M_2}	$\sigma_{M_{m-z}}$	σ_a	σ_{A_1}	σ_{A_2}	$\sigma_{A_{\theta-z}}$
All (742)	2.04 ± 0.07	1.97 ± 0.08	2.04 ± 0.07	1.33	1.36	0.93	0.93	0.304	0.300	0.231	0.231
Elliptical (163)	2.15 ± 0.18	3.69 ± 0.45	3.27 ± 0.41	1.24	1.33	0.90	0.90	0.184	0.309	0.176	0.155
Spirals (396)	2.01 ± 0.11	1.84 ± 0.10	1.98 ± 0.10	1.29	1.29	0.93	0.93	0.285	0.265	0.209	0.208
Lenticulars (158)	1.94 ± 0.13	2.51 ± 0.23	2.13 ± 0.16	1.21	1.17	0.79	0.79	0.218	0.271	0.170	0.166
Ellipiticals + spirals (559)	1.95 ± 0.08	1.80 ± 0.08	1.94 ± 0.80	1.34	1.31	0.94	0.94	0.312	0.286	0.232	0.231

sky being included in the sample. Averaging over the four possibilities yields $p = 2.06 \pm 0.12$. Again, scrutiny of the resulting dispersions in magnitude and logarithmic diameter unequivocally reinforces the phenomenological indications that $p \sim 2$. As earlier, dispersions of deviations from the $p = 1$ law are generally larger than those in the apparent quantities, and are in

TABLE 11
Effect of selection by region of sky

Portion of sky	Estimates of p			Dispersions in magnitude				Dispersions in log A			
	m–z	θ–z	m–θ–z	σ_m	σ_{M_1}	σ_{M_2}	$\sigma_{M_{m-z}}$	σ_a	σ_{A_1}	σ_{A_2}	$\sigma_{A_{\theta-z}}$
Whole sky (742)	2.04 ± 0.07	1.97 ± 0.08	2.04 ± 0.07	1.33	1.36	0.93	0.93	0.304	0.300	0.231	0.231
North celestial (525)	2.08 ± 0.09	2.13 ± 0.11	2.09 ± 0.09	1.30	1.35	0.90	0.90	0.290	0.306	0.226	0.226
North galactic (480)	2.19 ± 0.11	2.15 ± 0.14	2.19 ± 0.11	1.19	1.29	0.89	0.89	0.278	0.292	0.227	0.227
South celestial (217)	2.04 ± 0.14	1.67 ± 0.12	2.03 ± 0.14	1.34	1.36	0.94	0.94	0.333	0.285	0.243	0.240
South galactic (262)	1.90 ± 0.10	1.80 ± 0.10	1.89 ± 0.10	1.53	1.44	0.99	0.99	0.342	0.308	0.238	0.237

hardly any cases materially less than the latter. On the other hand, those from the $p = 2$ law are again quite materially less than those in the apparent quantities, and as earlier, in all cases less than those from the $p = 1$ law. Indeed, the raw comparison of dispersions of the residuals from the respective laws tends actually to overestimate the quality of the fit of the linear law, since for any theory of the form $m = f(z) + M$, $\sigma_m^2 = \sigma_f^2 + \sigma_M^2$, i.e., $\sigma_M^2 = \sigma_m^2 - \sigma_f^2$, the anticipated reduction in variance of the apparent magnitudes is thus σ_f^2, which is four times greater for the $p = 1$ law than for the $p = 2$ law.

Selection on both morphology and field of observation leads to further reduction in sample size, beyond which statistical investigation would appear likely to be moot. The results of this consequently virtually definitive refinement in sample selection, shown in Table 12, strikingly confirm the earlier conclusions. The average value of p for the three morphological types—elliptical, spiral, and lenticular—in the four indicated hemispheres of the sky, and as derived from the m–z relation, is 2.07 ± 0.24. In all cases the estimate is well within two standard deviations of the value $p = 2$. In the majority of cases, the dispersion in magnitude from the $p = 1$ line exceeds that in apparent magnitude and in no cases is it materially less, while in all cases the dispersion from the $p = 2$ line is considerably less than that in the apparent magnitudes, and within one percent of the minimal dispersion obtainable by a least-squares fit. The situation is generally quite similar for the θ–z and m–θ–z relations, except that the standard errors of estimate are greater and the results are consequently not quite as striking, although in precisely the same direction.

Due to the relatively small sample sizes, it would be inappropriate to make statistical analyses based on the postulate of radial spatial homogeneity, as is the V/V_m test described earlier, until a theoretical statistical procedure is available to deal with large local clusters such as Virgo and Fornax. These inevitably bias the spatial distribution of sufficiently small samples, and it would be improper, or at least statistically moot, simply to delete a priori local clusters designated in other than a functorial statistical fashion, i.e., by an objective procedure devoid of preconceived hypotheses as to the form of putative clusters.

On the other hand, the phenomenological viewpoint is primarily that of model-building, which is logically quite distinct from that of hypothesis testing. The lack of indication for the law $p = 1$ in the m–z–θ relation for low-redshift galaxies does not in itself imply that this law is statistically definitely unacceptable, for the degree of apparent magnitude truncation is considerable on the hypothesis that $p = 1$. However, the de Vaucouleurs data give a variety of further indications for the $p = 2$ law and counterindications for the $p = 1$ law.

TABLE 12
Effect of joint selection on morphological type and region of sky[a]

	North celestial	North galactic	South celestial	South galactic
Ellipticals				
	2.17 ± 0.20/3.59 ± 0.48/3.23 ± 0.47	2.32 ± 0.31/3.92 ± 0.81/3.17 ± 0.58	2.84 ± 0.76/5.07 ± 2.34/3.84 ± 1.38	2.07 ± 0.22/3.85 ± 0.56/4.47 ± 0.99
	1.38/1.51/0.95/0.95	1.18/1.34/0.94/0.93	0.78/0.94/0.71/0.70	1.24/1.29/0.83/0.83
	0.204/0.354/0.191/0.164	0.194/0.306/0.192/0.174	0.141/0.205/0.148/0.136	0.156/0.307/0.151/0.120
	(103)	(94)	(60)	(69)
Spirals				
	2.20 ± 0.15/2.06 ± 0.15/2.21 ± 0.15	2.19 ± 0.16/2.11 ± 0.17/2.18 ± 0.16	1.73 ± 0.14/1.53 ± 0.11/1.63 ± 0.12	1.79 ± 0.13/1.57 ± 0.11/1.73 ± 0.13
	1.18/1.30/0.89/0.88	1.15/1.24/0.88/0.88	1.46/1.24/0.94/0.93	1.54/1.36/1.00/0.99
	0.269/0.275/0.210/0.210	0.253/0.263/0.201/0.201	0.316/0.232/0.198/0.189	0.343/0.263/0.221/0.214
	(290)	(271)	(106)	(125)
Lenticulars				
	2.11 ± 0.17/2.92 ± 0.33/2.41 ± 0.22	2.06 ± 0.19/2.42 ± 0.28/2.17 ± 0.21	1.65 ± 0.21/1.82 ± 0.29/1.67 ± 0.22	1.73 ± 0.19/2.42 ± 0.39/1.95 ± 0.26
	1.11/1.18/0.72/0.72	1.06/1.09/0.72/0.72	1.37/1.11/0.89/0.88	1.35/1.16/0.89/0.88
	0.192/0.277/0.158/0.148	0.202/0.242/0.156/0.153	0.276/0.253/0.199/0.198	0.236/0.279/0.187/0.184
	(116)	(98)	(42)	(60)
Ellipticals and spirals				
	2.06 ± 0.10/2.00 ± 0.12/2.06 ± 0.10	2.14 ± 0.13/2.02 ± 0.14/2.14 ± 0.13	1.74 ± 0.12/1.40 ± 0.10/1.71 ± 0.12	1.76 ± 0.10/1.61 ± 0.10/1.74 ± 0.10
	1.32/1.36/0.93/0.93	1.20/1.28/0.91/0.91	1.32/1.14/0.88/0.88	1.56/1.34/0.97/0.96
	0.301/0.302/0.230/0.230	0.282/0.284/0.227/0.227	0.334/0.245/0.237/0.225	0.357/0.286/0.241/0.235
	(393)	(365)	(166)	(194)

[a] Within each classification: top line: maximum likelihood estimates together with corresponding dispersions, for the m–z, θ–z, and m–θ–z relations in order, for the indicated subsample of the basic sample of 742 galaxies with requisite data; middle line: dispersions in apparent magnitude, and in absolute magnitude for $p = 1$, $p = 2$, and $p =$ maximum-likelihood estimate, respectively; bottom line: the same for the logarithmic diameters. The number in parentheses is the sample size.

First, as is expected for a correct law, there is no significant trend with z in the absolute magnitudes based on the $p = 2$ law, while there is a pronounced trend for those based on the $p = 1$ law (cf. Figure 1). Second, the m–z relations of the galaxies in fixed redshift ranges, whether selected on brightness in bins containing fixed number of galaxies, or formed into a sample in their totality, are in very good agreement with the $p = 2$ law, but on the whole are no closer to the $p = 1$ law than they are to constancy. Representative results of this nature are given in Table 13, for all galaxies in

TABLE 13
Dispersions and mean magnitudes of bright low-redshift galaxies over assorted redshift ranges

Number of galaxies	Range in cz	σ_m	σ_{M_1}	σ_{M_2}	Mean m	Mean M_1	Mean M_2
NA	500–1100	0.99	1.00	0.97	11.57	24.44	18.01
251	500–1500	0.95	0.98	0.91	11.76	24.18	17.98
384	500–2000	0.90	0.98	0.85	11.91	23.89	17.91
159	600–1200	0.92	0.93	0.90	11.67	24.27	17.97
257	600–1600	0.93	0.96	0.90	11.82	24.06	17.95
368	600–2000	0.88	0.94	0.84	11.95	23.86	17.91

fixed redshift ranges, and Table 14, for relatively bright galaxies in bins. It is interesting to note that the spread in mean magnitude over different redshift ranges, for the same type of object, is much less for the $p = 2$ law than for the $p = 1$ law, the latter spread being on the whole no less than that in the apparent magnitudes. Thus in Table 13, the spread is 0.10 for the $p = 2$ absolute magnitudes, 0.38 for the apparent magnitudes, and 0.58 for the $p = 1$ absolute magnitudes. Similarly, for the fourth brightest galaxies in bins containing 10 galaxies each, these spreads are respectively 0.11, 0.61, and 0.50, over the redshift ranges considered in Table 14. Again, for the tenth brightest galaxies in bins containing 20 galaxies each, the ranges are 0.03, 0.33, and 0.34. The narrow spread of the $p = 2$ absolute magnitudes, and the generally undiminished spread of the $p = 1$ absolute magnitudes relative to that in the apparent magnitudes, is what would be expected on the $p = 2$ hypothesis, but is surprising on the $p = 1$ hypothesis. In particular, it is difficult to see how observational apparent magnitude truncation, admittedly a priori a conceivably significant factor, could result in such close agreement with the $p = 2$ law for such a variety of redshift ranges, bin sizes, and choices of relative galaxy brightness within each bin, if in fact p were equal to 1.

In any event, on the $p = 1$ hypothesis, any catalog complete out to a

TABLE 14
Dispersions and mean magnitudes of ranked galaxies in bins[a]

Sample criterion	Number of galaxies	Range in cz	σ_m	σ_{M_1}	σ_{M_2}	Mean approximate magnitude	Mean M_1	Mean M_2
First brightest, in groups of 10	50	500–3000	0.70	0.62	0.46	10.81	22.45	16.64
Second brightest, in groups of 10	50	500–3000	0.68	0.54	0.39	11.28	22.91	17.10
Third brightest, in groups of 10	50	500–3000	0.63	0.55	0.37	11.57	23.21	17.40
Fourth brightest, in groups of 10	38	500–2000	0.50	0.49	0.30	11.61	23.61	17.61
Fourth brightest, in groups of 10	50	500–3000	0.55	0.56	0.29	11.78	23.42	17.61
Fourth brightest, in groups of 10	61	500–5000	0.74	0.61	0.29	12.04	23.26	17.66
Fourth brightest, in groups of 10	67	500–∞	0.93	0.63	0.34	12.22	23.19	17.72
Fifth brightest, in groups of 10	50	500–3000	0.49	0.62	0.30	11.98	23.61	17.80
Tenth brightest, in groups of 20	20	500–2000	0.39	0.55	0.25	11.92	23.88	17.91
Tenth brightest in groups of 20	25	500–3000	0.43	0.61	0.24	12.05	23.70	17.88
Tenth brightest, in groups of 20	30	500–5000	0.61	0.67	0.24	12.25	23.54	17.90
Brightest of 11 at middle redshift of group of 11[b]	33	500–2000	0.63	0.54	0.44	10.62	22.60	16.61

[a] The term "rth brightest object in groups of s" refers to the procedure of arranging the source data in order of redshift, followed by subdivision into disjoint groups of size s (proceeding in the same order), followed finally by selection of the rth brightest object from each group.

[b] Again arranging objects in order of increasing redshift, those groups of 11 successive objects whose middle object is as bright as any in the group were picked out, and the sample formed from their middle objects.

limiting apparent magnitude $\bar{m}_{\text{app}}$ yields a fair sample in the range $z_1 < z < z_2$ if all objects intrinsically fainter than $\bar{M}_{p=1} = \bar{m}_{\text{app}} - 5 \log z_2$ are deleted. The absolute magnitudes $M_{p=1}$ for the resulting subsample should then exhibit a significant trend with z only if there is a corresponding luminosity evolution. The trend may be appropriately tested by comparing the mean of the subsample in the range $z_1 < z < z_3$ with that of the subsample in the range $z_3 < z < z_2$ by a t-test, z_3 being chosen so that the two subsamples have approximately the same size. (This test is "robust" for fairly large samples, meaning that no assumption of normality of the luminosity function is involved.) In fact, with the data of the de Vaucouleurs (1964), and the values $cz_1 = 500$ and $cz_2 = 2000$, and $\bar{m}$ in the range $12.5 \leqq \bar{m} \leqq 13$, the normal test statistic t is ~ 2.8, corresponding to a probability ~ 0.0025, indicative of a rapidity of evolution for nearby galaxies quite beyond normal expansion-theoretic conceptions. On the chronometric hypothesis, the corresponding t-value is 0.99, as is quite consistent with z-independence of the mean luminosity of bright galaxies. Additionally, if there is serious selection on luminosity for $cz \lesssim 2000$ in the de Vaucouleurs catalog, the population of deviations from mean magnitude of the subsample of galaxies in the vicinity of a fixed redshift should be noticeably redshift-dependent. A Smirnov two-sample test of these local model-independent luminosity functions in the vicinity of the redshifts $cz = 500$, 1000, 1500, 2000, and 2500, based on the groups of 20 galaxies nearest each redshift, reveals no significant differences between the distributions at the 5% level.

Continuing with the model-building discussion, a conceivable explanation for the phenomenological quadratic redshift–distance law within the expansion framework is the local superclustering proposed by G. de Vaucouleurs (1972). On the other hand, local superclustering would appear to involve significant radial spatial inhomogeneity. This is not at all confirmed by the Kolmogorov–Smirnov V/V_m test.

There is some evidence that the square redshift–distance law may persist at higher redshifts. The anomaly in the range $14 \leqq m \leqq 15$ reported by Rubin *et al.* (1973) is reduced from the significant level of five standard deviations to the insignificant one of two standard deviations if the linear law is replaced by a square one. The galaxies studied by Arakelyan *et al.* (1972) and described earlier are largely at higher redshifts and involve no known selection effects relatively favorable to a square law. However, the latter law decreases the dispersion in apparent magnitude, while the linear law increases it. The compilation of published redshifts of clusters of galaxies by Noonan (1973) exhibits a roughly linear $\log N$–$\log z$ relation for the 56 clusters in the range $z < 0.04$, of slope 1.46, which deviates only marginally from the square-law slope of 1.5 but considerably from the linear law slope of 3. The strong linearity shown by the Sandage sample of brightest cluster galaxies is, for reasons of small sample size, material model-dependence of the appropriate apertures of measurement for very large galaxies and unpublished selection criteria, not entirely conclusive as regards galaxies as a whole. In any event, the mean slope of ~ 1 for the (rather irregular) $\log N$–$\log z$ relation of the subsample in the range $z < 0.04$ is in better agreement with the square than with the linear law; and the former law does effect a significant ($\sim 50\%$) reduction in the dispersion of apparent magnitudes, in contrast with the absence of any reduction typically shown by the latter law for samples of other objects. It is interesting also that the $N(<z)$ relation of the Sandage sample, although highly irregular, varies roughly linearly with z rather than as $z^{3/2}$ for small redshifts, and even in the redshift interval $z < 0.04$ deviates considerably from the $N(<z)$ relation of the Noonan list of all galaxy clusters with published redshifts. Further analysis of the Sandage samples is given in the next section.

In summary, the data given on the de Vaucouleurs tape indicate that the hypothesis that $p = 1$ may be acceptable, with substantial emendations in the nature of superclustering, extreme breadth of the luminosity function, and the like. However, it is not phenomenologically indicated by the observations on low-redshift galaxies, which suggest rather the hypothesis that $p = 2$. This hypothesis leads to a narrow luminosity function, of breadth < 1 mag, and appears to be acceptable on the basis of all observable relations within the sample thus far examined.

9. The redshift–magnitude relation for Sandage's brightest cluster galaxies

In assessing the implications of this relation, it is necessary to bear the following circumstances in mind:

(a) The brightest cluster galaxy evolved in the work of Hubble, Humason, Mayall, and Sandage, as a means of confirmation and elaboration of the Hubble theory. Its independent status as a "standard candle" has not been established, and there is opposing evidence (cf. Abell, 1972; Peterson, 1970b; Zwicky, 1970).

(b) A sample sometimes cited as one of the main observational bases for the Hubble relation, given as Table 2 by Sandage (1972b), while undoubtedly of outstanding accuracy, appears to be of uncertain statistical uniformity. No objective criterion for a galaxy to be included in the sample has been published, nor indeed is it expressly claimed in the cited source that it is an appropriate sample for testing the redshift–magnitude relation. A superficial examination of the redshifts' ranges and numbers of galaxies indicates so clearly that it is in no sense an approximation to a complete sample out to a fixed limiting apparent magnitude, that a test of this via, e.g., the Schmidt V/V_{m} test may appear supererogatory. Rather the sample seems suitable for determination of the value of the deceleration parameter, on the basis of the prior hypothesis that a Friedmann model is valid. A sample well suited to this purpose may however be totally inappropriate for a test of this prior hypothesis.

(c) The statistical theory of the apparent uniformity of luminosity of the brightest galaxies in rich clusters, first clearly enunciated by Scott (1957), remains quite tenable (cf. Peterson, 1970b). If valid, the appropriateness of observations on bright cluster galaxies as a means of validating a theoretical hypothesis is further reduced.

(d) The small dispersion in absolute magnitude from the expansion-theoretic standpoint of the 41 galaxies studied by Sandage (1972b), of the order of 0.3, is in itself not at all a statistical verification of uniformity of their actual physical intrinsic luminosities, in view of the not necessarily random character of the sample involved, as well as neglect of evolutionary corrections. A concrete illustration of the ease with which such small dispersions may be attained by selection is afforded by the case of quasars, which in their totality are well known to fit expansion-theoretic redshift–magnitude curves with large dispersion, of the order of 1.7 mag in the case of the comprehensive list due to DeVeny *et al.* (1971). It is easy to select a subsample of 41 quasars which fit the curve with a dispersion of less than 0.3 mag, as is evident from a plot of the data (cf. below). Needless to write, no serious investigator would consider such a procedure valid; but the result of selection by a sufficiently refined physical criterion, or for a different statistical purpose such as the minimization of the variance of an estimate of a parameter (such as q_0) may be de facto virtually identical with this.

Finally, the Sandage (m, z) pairs cannot be corrected in any clear-cut fashion to obtain the magnitudes that should have been obtained if apertures appropriate to the chronometric theory had been used. In the case of the Peterson data, the precise apertures pertinent to the recorded magnitudes are given; such data are not available for the Sandage pairs. Indeed, the actual procedures employed in obtaining the final magnitudes from the observations are quite complex, and in particular: (a) the procedure employed is galaxy-dependent (in some cases a standard curve was used to correct to a presumed aperture, in others interpolation, etc.); (b) the standard aperture correction curve is presented in a fashion that implicitly assumes a certain relation between metric and isophotal diameters, which arises in the theory of Friedmann models, and is not valid in the chronometric theory; (c) the eye-fits used in processing data are difficult to treat in a statistically controlled fashion within the framework of an alternative theory.

The regrettable conclusion emerges that there is no entirely correct means to utilize the data presented in Table 2 of Sandage (1972b) in a statistically valid test of the chronometric hypothesis. More generally, it is doubtful whether data gathered for the efficient determination of q_0 on the hypothesis that a Friedmann model holds can legitimately and practically serve at the same time to test the latter, or an alternative, cosmological hypothesis. The actual dispersion from the chronometric theory for constant-luminosity objects of the Sandage (m, z) pairs necessarily differs by very little from the standard deviation of the difference between the theoretical curves

$$(5 \log z) - [2.5 \log z - 2.5 \log(1 + z)]$$

over the range of redshifts represented by the sample. With a *formal* correction for aperture based on the assumption that the correction procedure employed for the Peterson galaxies is somehow de facto valid for the Sandage galaxies, the actual dispersion of the 79 aperture-corrected (m, z) pairs in the redshift range $0.01 < z < 0.21$ from the chronometric curve is 0.74 mag. (In order to obtain a maximally homogeneous subsample it seemed appropriate to exclude five galaxies which are widely separated in redshift from the others in Sandage's total list of 84 brightest cluster galaxies.) The considerably smaller dispersion of 0.30 mag given by Sandage for the dispersion in expansion-theoretic absolute magnitude is explicable in terms of a variety of effects: an underestimate of the aperture correction, the Scott effect, inherent variability in the intrinsic luminosity of brightest cluster galaxies as indicated by Abell (1972), etc. The gross deviation from radial spatial uniformity in the sample is suggestive of a strong selection effect, which is borne out by the following analysis, and which alone is of magnitude quite sufficient to explain a chronometric dispersion of the value reported.

The cited data themselves indicate quite significant differences between the sample galaxies, exhibiting much variability and tending to support the statistical theory of the nature of the brightest cluster galaxy, as opposed to the theory that it is physically distinctive. For 22 of the galaxies, magnitudes are given at two or more apertures; from such data it is possible to estimate the Hubble radius of the galaxy according to the equation

$$m_1 - m_2 = -2.5 \log \left[\int_0^{r_1} rI(r/a)\, dr \middle/ \int_0^{r_2} rI(r/a)\, dr \right],$$

where m_1 and m_2 are the magnitudes, r_1 and r_2 are the radii at the galaxy corresponding to the given apertures θ_1 and θ_2, and I is the function earlier defined. The determination of radii r_j from the apertures θ_j together with the redshift z depends on the Hubble parameter H and also on the model. Taking as in Sandage's work $H = 50$ and using the simple Friedmann model with $q_0 = 1$ (the values of a are quite insensitive to the value of q_0) gives a fully specified equation for a which is readily solvable by successive approximations, using the largest and smallest values of the aperture listed by Sandage (1972b). The values of a which are thereby obtained are quite variable, ranging from below 0.05 kpc to above 45 kpc. While these extremes may well result from errors in magnitude observations, a large part of the variation must arise from other sources. The standard deviation of log a

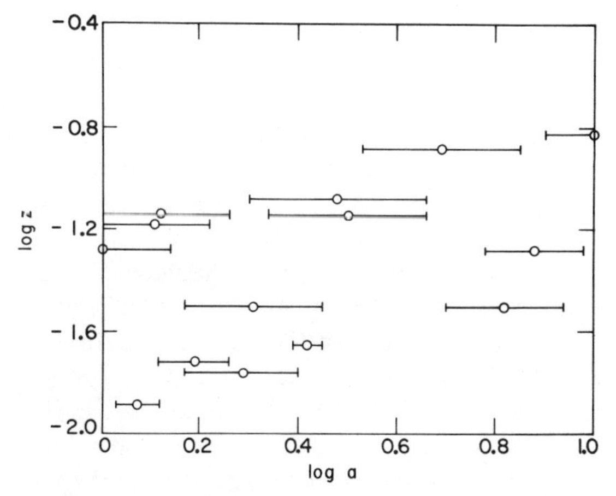

Figure 10 *The Hubble core radius–redshift relation for the brightest cluster galaxies with suitable data studied by Sandage* (1972b).

The error bars for individual galaxies are small relative to the differences in core radius between the galaxies, indicating nonuniformity in intrinsic size.

is 0.57; if four relatively extreme values of a are deleted, a quite connected group of values is obtained of dispersion 0.28 in log a. The actual (z, log a) pairs are shown in Figure 10. A change in the Hubble parameter would affect all values of a equally, and hence not alter the dispersion in log a. The change in the model from the chronometric to the Hubble one has the effect of multiplying the value of a determined on the basis of the chronometric theory by the ratio of the distances according to the respective models. For the redshift range involved here, this ratio does not differ effectively from const $\times$ $z^{1/2}$. This factor does not, however, produce a significant reduction in the dispersion in a.

Such dispersion should arise in major part from that in magnitude measurements, although this can be expected to be quite small in view of the accuracy of these measurements as described by Sandage (1972b). However, when the two apertures of measurements $\theta_{\min}$ and $\theta_{\max}$ are relatively close, the dispersion in magnitude measurement is particularly likely to be reflected in an apparent dispersion in log a. As a final means of estimating a conservative lower bound for $\sigma_{\log a}$, a subsample has been formed consisting of only those galaxies for which $\theta_{\max}/\theta_{\min} > 1.75$, and for which there is unambiguous data. For the resulting sample of seven galaxies, the dispersion in log a', where a' denotes the Hubble radius as determined on the basis of Hubble-law distances, is 0.18, which in view of the accuracy of the magnitudes can reasonably be attributed primarily to substantial variation in the size of the galaxies.

Such dispersion in log a therefore indicates considerable variability in an aspect which is strongly correlated with intrinsic luminosity. The relation between angular diameter and absolute luminosity has been treated in a comprehensive and precise study by Holmberg (1969), who finds (p. 326) that $M = -6.00 \log A + \text{const}$, within a dispersion of 0.40 mag. It follows that $\sigma_M \sim 6\sigma_{\log A}$, for any group of, e.g., elliptical galaxies (the constant is slightly type-dependent). Making the plausible assumption that the Hubble radius is sufficiently closely related to the diameter that a similar relation holds with A replaced by a, it follows that the order of magnitude estimate $\sigma_M \sim 6\sigma_{\log a}$ is likely to be valid. For the purpose of estimating the expected sample variance in absolute magnitude, the largest of the quoted standard deviations in log a may be the most relevant, but again, using the smallest figure, 0.18, for conservatism, it follows that a dispersion of the order of ~ 1.14 mag in absolute magnitude is to be expected for the galaxies in the group in question. This dispersion is of the same order as the actual dispersion of the data from the chronometric curve. There is no published basis for estimating the dispersion in log a for the entire group of 84 galaxies, but there is no apparent reason to doubt that the group for which data are given are representative at least of the correct order of magnitude. The smallness of the dispersion in expansion-theoretic absolute magnitude reported by Sandage suggests either an implicit selection effect, for which there are other indications, or an extraordinary physical uniqueness for brightest cluster galaxies which exempts them from even rough obedience to Holmberg's law. In the absence of any independent evidence for this exceptional behavior, the order of magnitude of the dispersion from the chronometric curve of the Sandage (m, z) pairs is no greater and indeed less than was to be expected from the dispersion in the Hubble radii a of that subsample for which Sandage has given data sufficient for its estimation.

Since the foregoing was written, Gunn and Oke (1975) have questioned procedures apparently involved in the earlier treatment of bright cluster galaxy samples. These include the attempted deletion of cD (supergiant) galaxies from the sample, despite the difficulty of recognizing them at large redshifts, and the subtraction of the background cluster luminosity, despite the difficulty of isolating that part of the luminosity due to the subject galaxy itself. It is evident that systematic deletion of particularly bright galaxies at larger redshifts would tend to bias the observational redshift–magnitude relation in the direction of increasing slope for the $m - \log z$ relation. It could also affect the distribution of intrinsic diameters of the sample. The ambiguity in the subtraction of the cluster background luminosity would not have this effect, but could simulate it in its impact on the dispersion in log a. It is evident that these considerations only enhance the general conclusions reached regarding the statistical admissibility of the Sandage sample, even if the second one may contribute significantly to the surprisingly large apparent dispersion in the absolute diameters of the sample galaxies. Finally, the difficult problem of a statistically sound definition of "cluster," earlier alluded to, is a further point to consider in assessing the significance of the bright cluster galaxy samples.

More recently, Sandage and Tammann (1975) have treated the m–z relation for ScI galaxies, again primarily for the estimation of q_0 and of intrinsic luminosities. Their reported result that $\partial m/\partial \log z \sim 5$ for these galaxies is obtained by combining two distinct samples, one consisting of classic ScI galaxies, and another of faint galaxies at generally much higher redshifts, whose identification as ScI galaxies is of quite another character and uncertain. The large difference between the average redshifts of the two samples results in an extreme sensitivity of the slope in the m–z relation to a possible substantial difference in average intrinsic luminosity between the two samples.

Indeed, the m–z relation of each of the separate samples is much better fitted by a line of slope 2.5 than one of slope 5. The respective dispersions in (a) apparent magnitude, (b) expansion-theoretic absolute magnitude ($q_0 = 1$), (c) chronometric absolute magnitude, are as follows. For the classic ScI galaxies (all unexceptionable data in Table 6 of Sandage and Tammann (1975), with $cz > 500$, a group of 22 galaxies), (a) 0.80, (b) 0.78, (c) 0.57. For the Sandage–Tammann sample of 60 galaxies among those in the list of 69 galaxies given as their Table 1: (a) 0.66, (b) 0.62, (c) 0.49. For the combined sample of 82 galaxies, the results are: (a) 1.55, (b) 0.67, (c) 0.90. These results imply that for the combined sample, the distribution of expansion-theoretic absolute magnitudes will show a pronounced cyclical trend. Since in addition to a major uncertainty as to the propriety of the classification of the faint galaxies as supergiants, the distribution of redshifts for the overall sample appears strongly nonrandom, the two samples of Sandage and Tammann are hardly consistent with the expansion hypothesis. Indeed, each sample itself deviates by $\gtrsim 4$ standard deviations from expectation, if the sample is assumed fair and the Hubble law is valid, as measured by the normalized reduced variance statistic introduced in Segal (1975). Each sample, however, is separately quite consistent with the chronometric hypothesis within $\lesssim 1$ standard deviation.

To summarize, the data for large or objectively designated galaxy samples are not at all phenomenologically indicative of a nonevolutionary expansion of the universe, but rather of the $m - z - \theta - N$ relations predicted by the chronometric redshift hypothesis. The data for samples that are small or may otherwise be less cogent statistically is generally similar, except for that on brightest cluster galaxies, which appears exceptional and equivocal. Greater definitiveness in the testing of the two hypotheses could probably best be achieved by additional observations or samples—randomized if necessary in specified fields—that are complete in redshift and out to specified limiting magnitudes. There is no model-independent reason to anticipate that such samples will be relatively more favorable to the expansion hypothesis, and indeed the sample of Colla *et al.* (1975) of this nature, published too recently to be detailed here, appears to be in satisfactory agreement with the chronometric $m - z$ and $N(<z)$ predictions, but poor agreement with the expansion-theoretic ones. In particular, the X-statistics earlier referred to (these are approximately normally distributed with zero mean and unit variance for a fair sample and correct theory), based on the raw data for the fifty-four radio ellipticals in the sample, are, respectively, -2.1 and 4.7; for the subsample of forty-four with $z < 0.05$, especially unlikely to be strongly affected by the observational magnitude cutoff, the values are -1.9 and 3.5. These represent formal probability ratios in favor of the chronometric theory of $> 10^5$ and 10^2, respectively.

10. Preliminary discussion of quasars

It has sometimes been asserted that quasar data have been disappointingly inapplicable to cosmological testing, by virtue of the large dispersion in their characteristics indicated by the data. The actual data, however, do not bear out this negative point of view regarding quasars, to the extent that model-independent analysis is possible. Moreover, from the standpoint of the chronometric hypothesis, their dispersion is quite moderate.

A priori one might expect that quasars would form an intrinsically more homogeneous class than galaxies. The cases in which there is some question as to whether a given object is a quasar are relatively few; the well-known variability in brightness is limited to a small fraction of the quasars, and introduces a dispersion in the magnitude too small to be of any significance in cosmological testing. Further, while relatively few quasars have been found as the result of statistically controlled observation, there are several important samples of this type, and the very heterogeneity of selection and of the telescopes involved in observation of the totality of known quasars should tend to prevent any strong bias from affecting the observations as a whole. Certainly, any selection effect on quasars has been fairly constant in the past six to seven years, for the m–z–$N(z)$ relation based on the ~ 70 quasars for which reliable data were available circa 1966 does not appear to differ appreciably from that based on the $\gtrsim 200$ quasars known today.

Actually, quasar data are in quite good agreement with the chronometric hypothesis, on the simplest possible model-independent hypotheses:

(1) spatial and temporal homogeneity (the latter meaning "no evolution," in particular);

(2) the quasars form a single luminosity class apart from a moderate roughly Gaussian random fluctuation. On the other hand, they are consistent with the expansion hypothesis only with the adjunction of model-dependent assumptions: (a) strong temporal evolution and spatial inhomogeneity; (b) a broad luminosity function, involving the existence of relatively large numbers of faint quasars for which there is little direct observational evidence.

Both assumptions (a) and (b) require the use of the observations themselves to determine the many parameters needed to specify fully the assumptions. The predictive power of the expansion hypothesis is thereby quite limited in regard to quasars, and its verification in the indicated sense would be possible with relatively arbitrary data. The recent work of Schmidt (1972a) details from the standpoint of the expansion hypothesis the parameters of the quasar population (cf. also the references to earlier work given there). The testing of the expansion hypothesis which is undertaken here is designed to parallel as closely as possible the tests applied to the chronometric hypothesis, in order to afford a fair and objective comparison, and so differ in format and, in part, in detail and in the quasar samples employed. The qualitative conclusions obtained are in no respect in disagreement with those of Schmidt (1972a), but his work stresses the determination of the quasar parameters on the assumption of the expansion hypothesis, while the present work is concerned rather with the comparison between the expansion hypothesis and the chronometric hypothesis. See also Longair and Scheuer (1967) for an analysis of the quasar m–z relation from a largely expansion-theoretic standpoint.

The major statistically controlled data regarding quasars are the lists by Schmidt (1968) of 3C quasars; that of Lynds and Wills (1972) of 4C quasars; of Braccesi *et al.* (1970) of radio-quiet quasars; and the summary material by Schmidt (1970) regarding radio-quiet quasars. Possibly subject to relevant selection effects, but so much larger in sample size as well as heterogeneous in selection as quite possibly to possess comparable statistical power, is the compilation by De Veny *et al.* (1971) of published data on quasars up to 1971. In addition there are recent lists of radio and radio-quiet quasars due to Sandage (1972c), of unspecified statistical applicability, and older lists such as that given by Burbridge (1967), the latter being of interest in relation to the question of the temporal stability of conclusions drawn from comprehensive heterogeneous lists.

These data have been treated in a systematic but simple statistical fashion. First, the redshift–magnitude relation has been compared with those predicted by the respective theories, on the assumption that the objects under consideration form essentially a single luminosity class with moderate dispersion. This assumption is confirmed by quantitative analysis, apart from the possibility of "temporal evolution" in the expansion model. In the vicinity of a fixed redshift, the dispersion in quasar apparent visual magnitudes as given by DeVeny *et al.* averages 0.8 mag, for redshifts $z > 0.2$ (for $z < 0.2$ the dispersion is ~ 1.3 if 3C 273 is excluded, the increase possibly being due to a slight degree of selection on luminosity, and/or the difficulty of distinguishing quasars from similar luminous objects, such as Seyfert galaxies). The precise situation as regards the DeVeny list and other quasar samples will be treated later, but an overall view of the situation is provided by Figures 11–13.

Figure 11 gives the standard deviation as a function of redshift for the apparent magnitudes of the quasars in the DeVeny list, at an approximately fixed redshift. Excluding quasars whose magnitudes are qualified as U, E, or

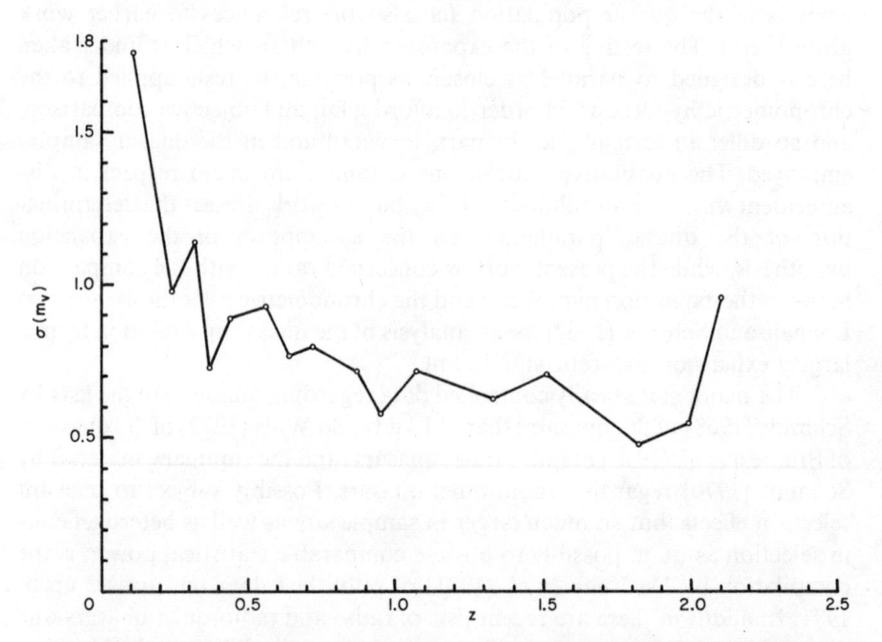

Figure 11 *Dispersion $\sigma(m_v)$ in quasar apparent magnitudes as a function of redshift.* All quasars in the list by DeVeny *et al.* (1971) having unqualified data, 158 in all, were divided into 16 groups of 10 quasars (8 in the last) of approximately equal redshift. The dispersion in each group of 10 quasars is plotted against the median redshift of the group (median taken as average of fifth and sixth largest redshifts in the group). For $z > 0.2$, the dispersion is comparable with that for bright galaxies on the same basis, and is not indicative of a material observational cutoff in apparent magnitude.

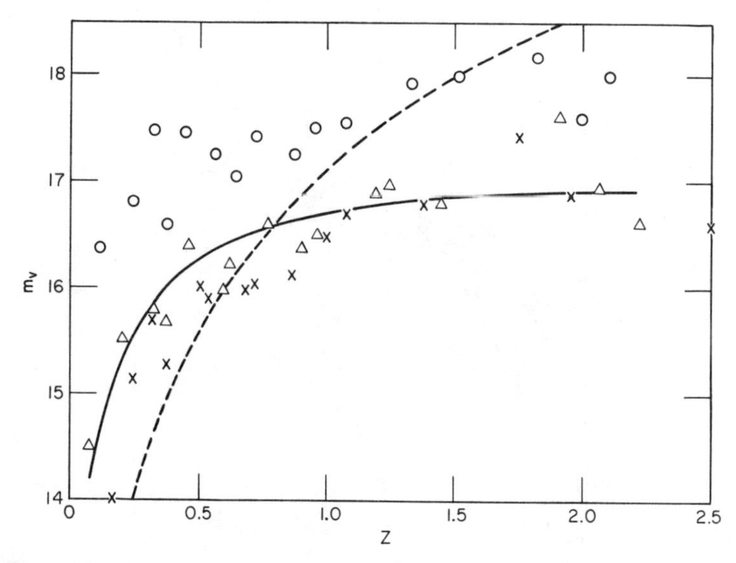

Figure 12 *Redshift–magnitude relations for quasars from* 16 *groups of* ~ 10 *quasars ordered by redshift.*

○, means of each group; △, second brightest quasar in each group; ×, first-brightest quasar in each group. The curves are theoretical constant intrinsic luminosity lines, *with constant adjusted from the second-brightest quasar observations*, on the following hypotheses: —, chronometric theory, $\sigma = 0.26$; - - -, expansion theory ($q_0 = 1$), $\sigma = 1.32$. For the first-brightest and mean quasars in each group, the respective theoretical curves must be correspondingly lowered and raised, and again provide an excellent fit in the case of the chronometric theory and a poor one in the case of the Hubble line.

P in this list, there are 158 quasars having measured redshifts and apparent magnitudes. These were divided into groups of size 10 in order of increasing redshift, the last group comprising eight quasars. The quantity $\sigma(m_v)$ plotted against redshift in Figure 11 is $[n^{-1} \sum (m - \bar{m})^2]^{0.5}$, where n is the number of quasars in the group, m denotes the apparent magnitude, and $\bar{m}$ the mean of the magnitudes in the group. The quasars in each group have slightly different redshifts, but on either the expansion or the chronometric hypothesis, these differences should contribute entirely marginal amounts to the dispersion of the group. Thus for example the widest redshift range is the last, which is $2.07 \leqq z \leqq 2.72$; the expansion-theoretical dispersion in apparent magnitude for the quasars in question, assuming they have the same intrinsic luminosity, is 0.19, and is still less on the chronometric hypothesis. The effect on the computed dispersion is likely to be much less; if the theoretical deviations are uncorrelated with the variations in intrinsic luminosity, the effect on the computed dispersion of 0.96 would be to reduce it to 0.94. Thus Figure 11 gives an effectively model-independent indication

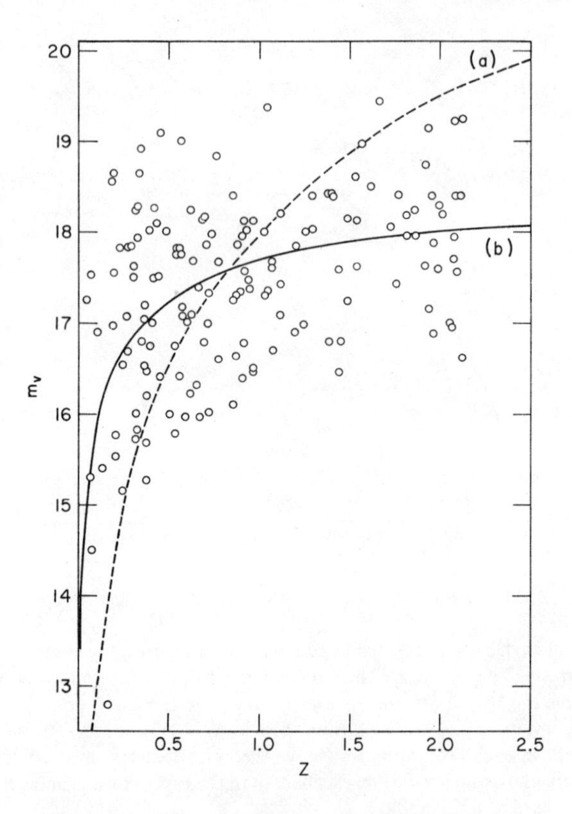

Figure 13 *The redshift–magnitude relation for quasars.*
○, All quasars with unqualified data in the list of DeVeny *et al.* (1971). Curves are the constant intrinsic luminosity curves for (a) the Friedmann model with $q_0 = 1$ and (b) the chronometric theory, with respective average luminosities fitted to the data. The chronometric curve closely approximates the mean position of the quasars at any given redshift; the Hubble curve is clearly systematically below the data for lower redshifts and above for higher redshifts.

of the observed dispersion in apparent magnitude for quasars at fixed redshift. (It would not be correct for a model in which magnitude varied rapidly with redshift, but any such model would be in gross contradiction with quasar observation, and need not be considered here.)

The lack of any pronounced downward trend in dispersion as a function of redshift, for redshifts > 0.2, is an indication of the absence of serious selection effects, as regards selection on luminosity. This indication is reinforced by Figure 12, which shows the redshift–magnitude relations for (a) the brightest quasar in each group; (b) the second brightest quasar in each group; (c) the mean quasar in each group (i.e., the mean magnitude plotted at the median redshift, which differs insignificantly from the geometric or arithmetic mean redshift here). If there were serious selection effects, the differences in magnitude between these respective observational curves would be likely to decrease with increasing redshift. In fact, these differences show no significant trend with redshift.

We shall treat in detail statistically controlled samples as well as the DeVeny list; but in view of the latter's model-independent apparent freedom from serious selection effects, it may reasonably be expected to afford a solid indication of the acceptability of the respective theoretical hypotheses, and/or their relative discrimination. In Figure 12, the best-fitting single-luminosity-class theoretical curves on either hypothesis, for the second-brightest quasar sample, are also plotted. It is evident that the expansion-hypothesis relation has a much greater dispersion from the observations than the chronometric relation; this is actually the case for the other two samples, as well. Figure 13 shows the totality of 158 quasars in the DeVeny list (having unquestioned magnitudes and redshifts), together with the theoretical chronometric and Hubble-theoretic curves of constant intrinsic luminosity, adjusted to the samples. The chronometric line appears virtually an optimal fit for a monotone increasing m–z relation (cf. also below).

The quantitative situation regarding this approach to the redshift–magnitude analysis of the DeVeny quasars is summarized in the following tables. Table 15 lists: (1) the redshift rank of groups of quasars, each having

TABLE 15
Redshift–magnitude observational and theoretical data for quasars in groups ordered by redshift

1	2	3	4	5	6	7	8	9	10	11
1	0.11	16.38	1.76	Mk 205	0.070	14.50	−3.16	0.06	0.158	12.80
2	0.24	16.79	0.98	PKS 2135 − 14	0.200	15.53	−1.91	0.08	0.240	15.15
3	0.32	17.47	1.14	PKS 2251 + 11	0.323	15.82	−1.16	−0.05	0.311	15.72
4	0.37	16.60	0.73	Ton 202	0.366	15.68	−0.75	−0.29	0.371	15.28
5	0.44	17.46	0.89	PHL 658	0.450	16.40	−1.02	0.27	0.501	15.99
6	0.56	17.25	0.93	3C 345	0.594	15.96	0.02	−0.37	0.530	15.78
7	0.64	17.05	0.77	MSH 03 − 19	0.614	16.22	−0.17	−0.13	0.677	15.97
8	0.72	17.43	0.80	3C 175	0.768	16.60	−0.06	0.11	0.720	16.02
9	0.87	17.25	0.72	4C −03.79	0.901	16.38	0.50	−0.21	0.859	16.10
10	0.95	17.49	0.58	3C 94	0.962	16.49	0.54	−0.14	0.980	16.47
11	1.07	17.55	0.72	PKS 1127 − 14	1.187	16.90	0.58	0.16	1.070	16.70
12	1.33	17.92	0.63	BSO 1	1.241	16.98	0.60	0.22	1.375	16.79
13	1.51	17.99	0.91	3C 298	1.439	16.79	1.11	−0.04	1.434	16.46
14	1.82	18.18	0.48	PHL 1222	1.910	17.63	0.89	0.69	1.750	17.43
15	1.99	17.61	0.55	PHL 1305	2.064	16.96	1.72	−0.01	1.955	16.88
16	2.10	18.00	0.96	PHL 8462	2.224	16.63	2.22	−0.37	2.720	16.60

10 members except group 16, having eight; (2) the median redshift of the group; (3) the arithmetic mean of the magnitudes of the group; (4) the standard deviation of these magnitudes about their mean (this is the conventional, and hence biased statistic; the figures should be increased by 5% for an unbiased estimate); (5) a conventional name for the second brightest object in each group; (6) the redshift z of this object; (7) the magnitude m of the object; (8) the Hubble deviation, i.e., the difference $m - 5 \log z + c$, the constant c being chosen so that the average difference vanishes; (9) the chronometric deviation, i.e., the difference

$$m - 2.5 \log z + 2.5 \log(1 + z) + c',$$

the constant c' being chosen so that the average of these differences vanishes; (10) the redshift of the brightest object in each group; (11) the magnitude of this object. No essential improvement in the dispersion of the expansion theory would be expected from the use of a Friedmann model in place of the simple Euclidean version; explicit computations were made with the Einstein–de-Sitter model, which actually produced an increase in dispersion. The main conclusion to emerge, that the dispersion of the expansion theory is of the order of 3 or more times greater than that of the chronometric theory, as regards the m–z relation for the indicated model-independent objects, should be unaffected by the use of another Friedmann model with a value of q_0 in the range generally considered realistic.

The actual dispersions are summarized in Table 16. Under "sample" is

TABLE 16

Dispersions and means for redshift–magnitude relations of quasars in groups ordered by redshift

Data	Theory: Chronometric	Theory: Expansion[b]	Sample
Second brightest in group	$\sigma = 0.26$	$\sigma = 1.32$	$\sigma = 0.71$ $\bar{m} = 16.34$
Group means	$\sigma = 0.29$	$\sigma = 1.34$	$\sigma = 0.49$ $\bar{m} = 17.40$
First brightest in group	$\sigma = 0.29^a$, 0.60	$\sigma = 1.02^a$, 0.99	$\sigma = 0.60^a$, 1.01 $\bar{m} = 16.01$

[a] Excluding 3C 273.
[b] Expansion theory: Friedmann model, $q_0 = 1$.

given the dispersion in the apparent magnitudes of the objects selected from each group, and also the mean apparent magnitude. There is clearly no qualitative difference between the results of using the second, first, or mean quasar in each group.

11. The N–z relation for quasars

The foregoing analysis has been primarily in the nature of model formation. It does not precisely constitute testing of hypotheses as to the correct model. It indicates that the chronometric hypothesis is quite satisfactory as regards the m–z relation for quasars, but that the expansion hypothesis appears to require emendation to be consistent with this relation. However, it remains to check out both hypotheses on other, statistically more controlled quasar samples; to test both hypotheses against other observational relations; to examine the modifications of these tests when the original hypotheses are augmented by auxiliary ones. In particular, the auxiliary hypotheses of strong temporal evolution in the properties of quasars, and the related one that they form a quite broad rather than roughly single intrinsic luminosity class, have been proposed in connection with the expansion hypothesis, and should be considered.

We continue now with our description of the statistical procedure for tests based on quasars. The next relation considered is that between $N(z) = N(< z)$, the number of quasars at redshifts $< z$, and z. In order to arrive at a definite relation, our earlier assumption to the effect that the objects form a single luminosity class with moderate dispersion must be augmented by an assumption as to their spatial distribution; it will be assumed that this is approximately uniform, i.e., the number of quasars in a given volume of space is generally proportional to the volume. This assumption will again be confirmed by the quantitative analysis of the quasar observations on the chronometric hypothesis, but in any event it is the physically most reasonable and mathematically simplest a priori distribution from which to commence model building or preliminary testing (as in the work of Schmidt, 1968, based on the expansion hypothesis). As a means of obtaining a preliminary overall indication, we again turn to comprehensive, independently compiled data.

As is well known, there are few quasars known of redshift > 2.5, and it is reasonable to anticipate that selection effects are present at somewhat lower redshift. Thus it would be surprising if the DeVeny quasar list approximated a random sample of quasars out to redshift 2.9, nearly the largest known redshift; but as z decreases, it becomes increasingly likely that those at lower redshifts approximate a random sample out to redshift z. We shall apply the Kolmogorov–Smirnov test for the comparison of the observed with the theoretical redshift distributions to various redshift intervals of this type. In addition, the redshift intervals (0.5, 1.0) and (1, 2) will be considered, as a check on the possible influence of lower redshifts, where anomalies may arise from classification difficulties. The results are that the chronometric hypothesis is accepted, at notably high levels of probability in most cases, while the expansion hypothesis is rejected in virtually all cases, at conventional significance levels.

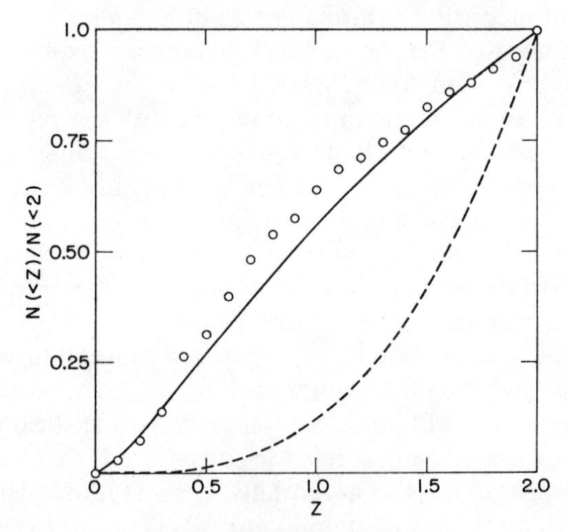

Figure 14 *The N–z relation for quasars in the redshift range* $0 < z < 2$.
O, cumulative fraction of quasars in the list by DeVeny *et al.* (1971) having unquestioned data, in the cited redshift range (146 quasars in all); —, expected distribution on the chronometric hypothesis; - - -, expected distribution on the Hubble theory for a random sample. Theoretical assumptions: spatial and temporal homogeneity. The quite satisfactory agreement of the chronometric prediction with the observations serves to confirm both the chronometric hypothesis and the assumption that the sample is fair. On the expansion hypothesis, this assumption is open to question due to the faintness expected for objects at larger redshift, but Figures 15 and 16 do not reveal any material improvement in the fit of the expansion prediction to the observations when consideration is limited to quasars in restricted redshift ranges.

Figures 14 and 15 show the observed and theoretical fractions of quasars of redshift bounded by a given value z. This form of presentation, as opposed to that of the differential fractions, has the advantage of being readily subject to statistical analysis. The Kolmogorov–Smirnov statistic, which is the maximum of the (absolute values of the) deviations of the observed and theoretical fractions, has a distribution which is independent of the true redshift distribution (this is strictly true only for continuous distributions, but in all events it yields an upper bound on the probabilities,

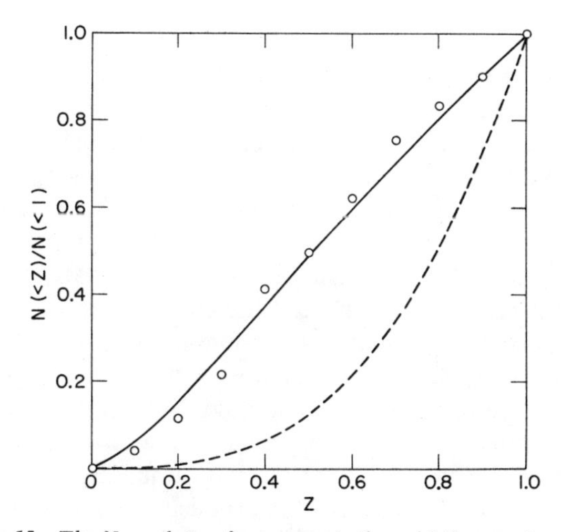

Figure 15 *The N–z relation for quasars in the redshift range* $0 < z < 1$.
○, cumulative fraction of quasars in the list by DeVeny *et al.* (1971) having unquestioned data (98 quasars in the cited range); —, expected distribution on the chronometric hypothesis; - - -, expected distribution on the Hubble theory for a random sample. Theoretical assumptions: spatial and temporal homogeneity. See the comment on Figure 14.

and there is no reason to anticipate a discontinuous distribution in the present case). It is clear visually that the observed distribution is well fitted, in all redshift intervals $(0, a)$, for $a \leqq 2$, by the theoretical chronometric distribution, while its deviation from the expansion-theoretical distribution is even greater than that between the two theoretical distributions. The situation is the same when the lower redshifts are excluded as in the intervals (0.5, 1) and (1, 2).

These results do not signalize rejection of the expansion hypothesis as such, but only its conjunction with the auxiliary hypothesis indicated. Spatial uniformity of the distribution may be in question, and in Schmidt (1968) and numerous analyses of quasar distributions technically along different lines from the present one, although related in general idea, the conclusion has been reached that it does not hold, if the expansion-theoretic hypothesis is correct. A fuller comparison with these earlier developments will be given later in connection with the chronometric results for the quasar samples treated by Schmidt *et al.*, but a preliminary indication of the extent to which the general hypothesis of z-dependence of the spatial distribution of quasars (mathematically equivalent to temporal evolution in the Friedmann model framework) may be accepted by the data may be obtained as follows.

Accepting provisionally this hypothesis, it would be anticipated that over relatively small redshift intervals the redshift distribution should conform to a nonevolutionary model. In fact, over very few intervals beginning at $z = 0$ is this the case; apart from the decrease in sample size involved, which limits the statistical significance of the conclusions, and considering the decrease in z-range, the expansion hypothesis fits about as poorly over the intervals $(0, a)$ for small a as for large values of a. This result might be explained on the basis of local anomalies, in particular the difficulty of discriminating between quasars and Seyfert and N-galaxies, and other large redshift emissionline objects. Virtually all such known galaxies having quasarlike features are at redshifts $\leqq 0.3$. The redshift intervals $(0.3, a)$ for values of a somewhat greater than 0.3 should therefore be substantially free from local anomalies and contaminations by nonquasars, and at the same time represent regions of space sufficiently close (on the expansion hypothesis) that selection effects on luminosity should be minimal. These intervals would therefore appear a priori as probably the most favorable ones for showing the approximate spatial uniformity of quasars over small redshift intervals, say ~ 0.2, on the basis of the expansion hypothesis.

In fact, the exclusion of the initial redshift interval (0, 0.3) does not significantly improve the fit of the expansion-theoretical distribution to the observations, over shorter redshift intervals, as shown by Table 17 for the intervals $(0.3, 0.3 + b)$ for $b = 0.2, 0.3, 0.4$. In all cases, the chronometric curve agrees with the observational line within quite probable random fluctuations, while in most cases the deviation of the expansion curve from the observational line is significant at conservative statistical levels.

TABLE 17
Kolmogoroff–Smirnov tests of the $N(z)$ relation for quasars

Redshift interval	Number in sample	Chronometric D^a	Chronometric probability^b	Hubble D^*	Hubble probability
0–2.0	146	0.09	0.19	0.42	10^{-22}
0–1.0	98	0.05	~ 1.00	0.38	2×10^{-12}
0–0.5	48	0.10	0.76	0.32	2×10^{-4}
0.5–1.0	50	0.12	0.47	0.29	5×10^{-4}
1.0–2.0	48	0.11	0.36	0.22	0.02
0–0.3	21	0.12	~ 1.00	0.28	0.07
0.3–0.5	27	0.28	0.09	0.35	0.003
0.3–0.6	40	0.16	0.26	0.30	0.002
0.3–0.7	52	0.12	0.45	0.27	0.001

[a] D, the Kolmogoroff–Smirnov statistic, is the maximum of absolute value of difference between observed and theoretical cumulative frequency functions.

[b] Probability is that of a D as large as that observed. Data from DeVeny *et al.* (1971), and comprise all quasars listed of unquestioned redshift and magnitude.

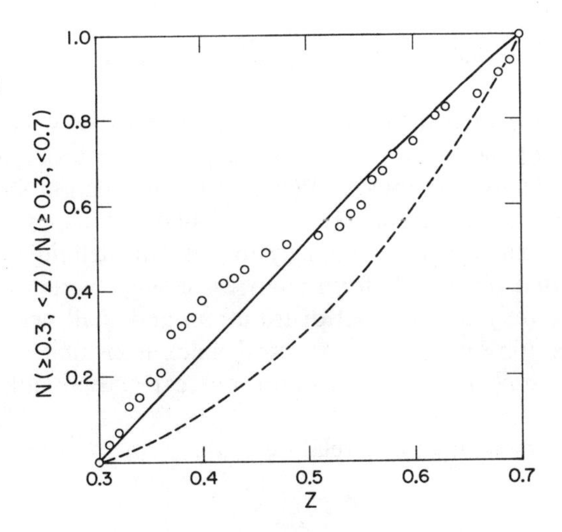

Figure 16 *The N–z relation for quasars in the redshift range* $0.3 < z < 0.7$. The basis here is the same as in Figures 14 and 15. Compare the comment on Figure 14.

The entire interval $0.3 < z < 0.7$ is shown in Figure 16. The tentative indication seems virtually inescapable that on the expansion hypothesis, the temporal evolution must be so rapid that even over redshift intervals of the order of 0.2, the expansion cannot be regarded as approximately stationary. This indication naturally recalls the related indication provided by the apparent spatial distribution of the Peterson galaxies described earlier. At the same time, the chronometric hypothesis fits the data remarkably well; even on a correct hypothesis, there might well be some nontrivial interval for the variate in question within which the sample distribution differs significantly from the population distribution, but no such interval of order $\geqq 0.1$ is apparent for the present observations in relation to the chronometric hypothesis.

Table 17 summarizes the results of Kolmogorov–Smirnov tests in the indicated redshift intervals. In order of magnitude, the probabilities on the expansion hypothesis that the observed deviations could arise by chance seem much smaller than might have been anticipated prior to the present analysis. Those for the chronometric hypothesis are, however, correspondingly remarkably large; it would be improbable for them to be much larger, even granting the validity of the theory. The indication from this latter circumstance is that the DeVeny list is rather more representative out to redshifts ~ 2 than one had any right to expect. The speed and thoroughness of observational quasar work during the past decade has perhaps been underestimated.

12. The apparent magnitude distribution for quasars

There is one final distribution of quasar statistics which is generally taken account of and logical to treat in the present context, that of apparent magnitudes. Ideally, the joint z–$N(<z)$–$N(<m)$ relation should be considered; however, the present state of the available statistics and also of the statistical art is such that it is unlikely to yield any definite useful information beyond that obtainable from analyses of single variates. We consider here therefore only the $N(<m)$ distribution, and shall neglect secondary effects such as deviation of the spectral index from unity (cf. the earlier treatment of the N–S relation for radio sources), and possible intergalactic absorption.

For a theory with an m–z relation of the form

$$m = f(z) + c,$$

and for a single luminosity class (i.e., fixed c), the apparent magnitude cumulative probability $P(<m)$ can be derived from the form of the function f, together with the distribution of z implied by the underlying geometry. If c itself is statistically distributed, then m becomes the sum of the variates $u = f(z)$ and $v = c$. Assuming that the luminosity function is independent of z, the distribution of m is then the convolution of the respective distribution functions for u and v.

There is no compelling reason to anticipate a normal distribution for the quasar intrinsic luminosities, but the chronometric luminosities are reasonably well approximated by this distribution; cf. Table 18, based on the 158 quasars in the DeVeny list. This is not the case for the expansion-theoretic luminosities, but the well-known law $N(m) \propto 10^{-0.6(\bar{m}-m)}$ for the distribution of apparent magnitudes below a fixed limiting magnitude is independent of the luminosity function (cf. Longair and Rees, 1972). Since in fact the normal distribution is approximated for computational purposes by a linear combination of delta-functions, and the theoretical $N(m)$ relation is not very

TABLE 18
Distribution of chronometric intrinsic luminosities

Deviation m from mean magnitude	Observed frequency	Normal law frequency, $\sigma = 0.9$ mag
$1.5 < \Delta m$	0.06	0.05
$0.5 < \Delta m \leq 1.5$	0.21	0.24
$-0.5 \leq \Delta m \leq 0.5$	0.43	0.42
$-1.5 \leq \Delta m < -0.5$	0.28	0.24
$\Delta m < -1.5$	0.02	0.05

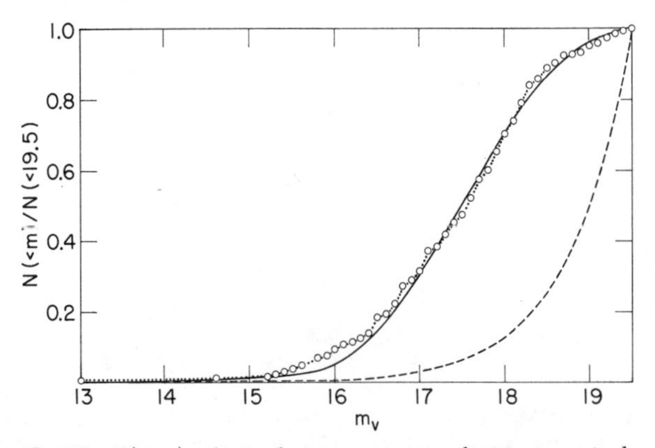

Figure 17 *The* $N(<m)$ *relation for quasars up to a limiting magnitude of* 19.5.
○, observational points; —, theoretical line for the chronometric theory, which depends on the observational luminosity function derived from the list of DeVeny *et al.*; - - -, theoretical line for the Hubble theory and is independent of the luminosity function. In either case it is postulated that the sample is a random subsample of a complete sample. From an expansion-theoretic standpoint this may be questioned, but the restriction to the fairly conservative limiting magnitude of 18 in Figure 18 still does not bring the expansion prediction into agreement with observation.

sensitive to the precise form of the luminosity function, the use of the normal law in the chronometric case does not differ significantly from the use of the empirical distribution function.

Accordingly, the convolution of the chronometric-theoretical constant luminosity curve with a Gaussian of dispersion 0.9 mag, approximately that found for the DeVeny list, has been computed and is compared in Figures 17 and 18 with the Hubble-theoretical curve and the observational apparent magnitude distribution for the same list, out to prescribed limiting magnitudes. The limiting magnitude of 19.5 in Figure 17 includes virtually all quasars on the list. It is evident that the chronometric curve and the observations are in extremely good agreement. On the other hand, the fit of the expansion-theoretic curve from the observations is quite poor; there is a notable deficiency of faint quasars, from the expansion-theoretic standpoint. In defense of the expansion theory, it might be argued that the sample is not known to be random, and that selection on luminosity might well be an important factor. That this is not very significantly the case is indicated by Figure 18, in which the limiting magnitude is 18, only slightly fainter than the modal quasar magnitude for a variety of samples. On the other hand, while the data clearly supply confirmation of the chronometric hypothesis, in the absence of model-independent methods of estimating selection effects this apparent rejection of the (nonevolutionary) expansion hypothesis should be regarded as tentative. Definitive tests should be sought through the use of statistically controlled data, samples of which are later discussed.

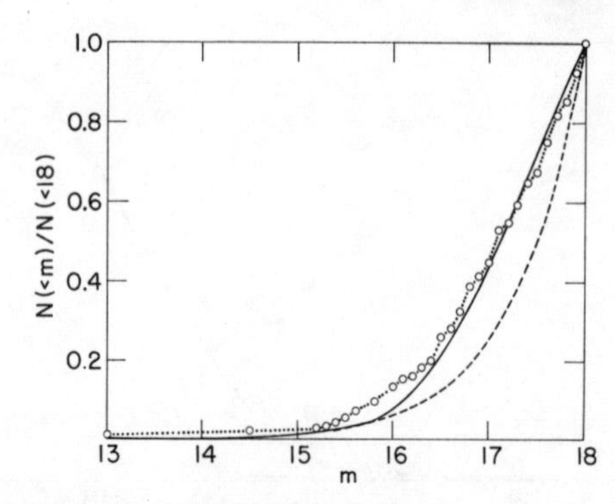

Figure 18 *The* $N(<m)$ *relation for quasars up to a limiting magnitude of* 18. The basis here is the same as in Figure 17. Compare the comment regarding that figure.

13. The redshift–luminosity relation for quasars

In addition to comprehensive lists such as the DeVeny list used earlier, there are the following selective lists: (1) Schmidt (1968), 3C sources; (2) Lynds and Wills (1972), 4C sources; (3) Braccesi *et al.* (1970), optically selected quasars. The Schmidt and Lynds–Wills lists include radio luminosities, and the Braccesi list includes infrared luminosities. The results of testing the *m–z* relation are indicated in Table 19. It is evident that the chronometric hypothesis is generally much more satisfactory in relation to the cited data. The chronometric dispersions are reassuringly uniform, ranging from $\lesssim 1$ mag for unselected quasars down as brighter quasars are selected. The Schmidt (1968) and Lynds and Wills (1972) lists being primarily radio-selected, it was to be expected that for a correct theory, the dispersions in the apparent radio luminosities should be relatively small, as they are relative to the dispersions in optical magnitudes. The Braccesi list being optically selected, it was similarly to be expected that it would show relatively small dispersions in apparent optical luminosities, as is the case. These appear to be the statistically best controlled data available, and are consistent in yielding dispersions of the order of 0.8 for samples which are complete, but include quasars of fairly low luminosities, in the model-independent sense of luminosity relative to other sample members of approximately equal redshift.

On the other hand, the expansion-theoretic dispersions are quite variable; in all cases higher than the chronometric and sample dispersions, for the most part quite substantially so. In model-independent terms, 3C 273 is exceptional in that it is more than 4 mag brighter than the average of the six quasars having the most similar redshifts (three greater, and three less

TABLE 19
The redshift–magnitude relation for quasar samples

Sample	Sample Size	Dispersions (magnitudes) Chronometric	Hubble	Sample
DeVeny, all objects having unquestioned z and V	158	0.95	1.67	1.02
DeVeny, replacing magnitude by average magnitude of seven quasars of nearest redshift (three above, three below)	152	0.32	1.23	0.52
DeVeny, locally brightest[a] 20% ("local brightness" measured by excess of magnitude above average of those of the six quasars of nearest redshift)	32	0.31	1.08	0.65
DeVeny, locally brightest[a] 10%	16	0.28	1.08	0.54
Schmidt, complete 3C sample	32[a] 33	0.80[a] 0.97	1.12[a] 1.12	0.88[a] 1.16
Lynds–Wills, complete 4C sample	30	0.89	1.32	0.99
Braccesi, all with unquestioned redshifts	27	0.79	2.28	0.58
Schmidt, complete 3C sample, radio magnitudes[b]	32[a] 33	0.72[a] 0.73	1.27[a] 1.27	0.67[a] 0.75
Lynds–Wills, complete 4C sample, radio magnitudes[b]	30	0.80	1.54	0.72
Braccesi, infrared magnitudes	27	0.88	2.32	0.64

[a] Excluding 3C 273.
[b] Reported values as corrected were converted to the Pogson scale.

than that of 3C 273; cf. below); for no other quasar is this difference as much as 2 mag. Consequently, it may be excluded on a rational statistical basis, and it seems more illuminating to do so. The overall indication from Table 19 is that the chronometric theory generally provides a distinctly and uniformly better fit.

Equally statistically significant with the comparison between the dispersions of the respective theories is the comparison between their dispersions and that from the sample mean. In all cases except the complete samples, the deviations from the chronometric theory have a lesser dispersion than those from the sample mean, as would be expected from a correct theory. In the event of a large dispersion in intrinsic luminosity, chance fluctuations could produce a slightly larger dispersion from a correct theory than from the sample mean, particularly when there is strong selection on apparent luminosity, as in the case of a complete sample, but it is extremely unlikely to produce a substantially larger dispersion. This applies to the much greater dispersion from the expansion theory than from the sample mean, in all cases except that of the Schmidt sample. This dispersion from the Hubble line is too large to be consistent at any acceptable probability level with a small dispersion in the intrinsic luminosities. On the other hand, if the latter dispersion is large, the relatively small dispersion of the magnitudes from the constant sample mean is then extremely improbable, in view of the considerable variation in 5 log z over the redshift range in question.

The conclusion seems inescapable that these data are in conflict with the expansion hypothesis, unless it be assumed that the intrinsic luminosities do not form a z-independent population. The need for this assumption, which from the standpoint of the expansion-theoretic hypothesis is naturally regarded as luminosity evolution, seems not seriously disputed by proponents of the expansion hypothesis, and need not be belabored here. It seems necessary to stress, however, that the assumption virtually eliminates the predictive power of the expansion theory as regards the luminosities of large-redshift objects. No such assumption is required for the chronometric hypothesis, which has quite significant predictive power. For example, the expansion hypothesis carries no implication regarding the probable magnitude of quasars which may be observed at redshifts ~ 3.5, which is essentially different in principle from that obtainable by simple extrapolation of the empirical m–z relation. According to the chronometric hypothesis in totally uncorrected form (with $\alpha = 1$), $m = 2.5 \log z - 2.5 \log(1 + z) + c$, then fitting the mean intrinsic luminosity index c to the DeVeny data, $c \sim 18.4$, yields the results $m \sim 18.1$. Interestingly, the quasar OH 471 reported by Carswell and Strittmatter (1973) of redshift 3.4 and the quasar OQ 172 reported by Wampler *et al.* (1973) of redshift 3.5 are approximately of this apparent magnitude, although their intrinsic luminosities are quite unprecedented from the expansion-theoretic standpoint, and would further exacerbate the problem of the quasar energy mechanism in this theory.

As a final aspect of the quasar m–z relation, we shall essay a test of the hypothesis that the *bright* quasars follow the chronometric and/or the expansion-theoretic law, employing an entirely model-independent definition of "bright," which also avoids the necessity for grouping quasars as has done earlier. This at the same time affords a model-independent estimate of the quasar dispersion in intrinsic luminosity. We shall define "bright" as bright relative to quasars at approximately equal redshift; more specifically, for any quasar we shall define the "excess brightness" Δm as the excess of the average magnitude of the six quasars obtained by selecting from the DeVeny list the three of nearest larger redshift and the three of nearest smaller redshift, over the magnitude of the quasar in question. We shall then consider the redshift–magnitude relation for the 10 and 20% of the sample for which this relative brightness is greatest. As earlier noted, the quasar 3C 273 is clearly exceptional in its relation to the distribution of excess brightnesses, having $\Delta m > 4$, while for all other quasars $\Delta m < 2$; accordingly, it will be excluded from these samples, which will be called for brevity the brightest tenth and fifth.

The results are included in Table 19 and shown in part in Figure 19. The brightest tenth of the DeVeny list has a chronometric dispersion of 0.28 mag, entirely without correction, less than that of the best samples of bright cluster galaxies. The dispersion of the same quasars from the Hubble

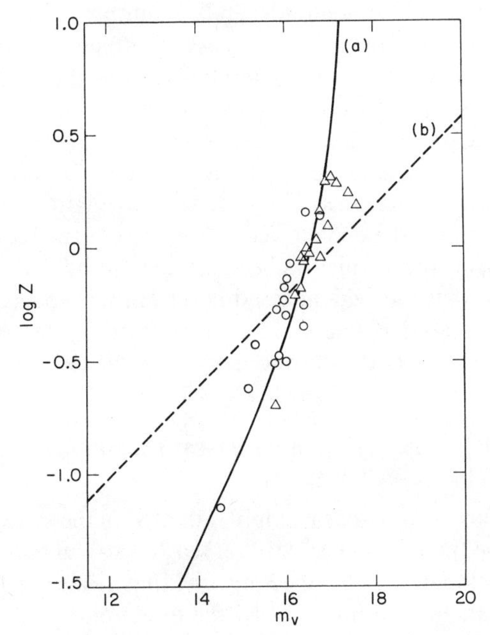

Figure 19 *The redshift–magnitude relation for the locally brightest fifth of the quasars in the list of DeVeny et al.*

○, quasars in the locally brightest tenth; △, quasars in the second-brightest tenth. Curves are best-fitting theoretical constant-intrinsic-luminosity curves for: (a) chronometric theory; (b) Hubble theory. As is representative for brighter quasars, selected in any fashion not making explicit use of a particular model, the dispersion from the Hubble line is more than three times that from the chronometric curve.

line is, however, 1.08 mag, i.e., even these bright quasars bear virtually no significant relation to the Hubble line. For the brightest fifth, consisting of 32 quasars, the dispersions are respectively 0.31 and 1.08 mag. (It should be noted that the sample dispersion is 0.65, less than that from the Hubble line.) These dispersions are fully comparable with those obtained for most samples of bright cluster galaxies after correction for color and galactic absorption. This suggests that the relatively bright quasars form a "standard candle" at least to the same extent as brightest cluster galaxies may do so; the standardization is further augmentable by selection on radio spectral indices, as proposed by Setti and Woltjer (1973) (cf. below).

It should perhaps be noted that work of Bahcall and Hills (1973), which appeared after this manuscript was largely complete, is directed toward establishing that the "brightest" quasars follow the Hubble law. The definition of "brightest" is in part model-dependent, and only seven quasars are included in the final sample found to have a dispersion of 0.3 from the Hubble line. This dispersion is no less than that from the chronometric prediction of the present model-independent samples of size 16 and 32.

Finally, we mention that various quasar samples of undesignated selection criteria show the same m–z relation behavior as the ones just discussed. The largest such list, apart from the DeVeny list, is that of Sandage (1972c). The chronometric dispersion is markedly less than the Hubble line dispersion for the complete Sandage list, the subsample of 15 radio-quiet quasars, and also for radio luminosities. The analysis of the latter involves transformation of the model-dependent data listed by Sandage back to their presumed empirical form; this has been carried out by J. F. Nicoll. Nicoll's results also show that the evident trend in the Hubble absolute radio luminosity with z, remarked by Sandage and ascribed by him to selection, is entirely accounted for by the chronometric theory; see also Section 17.

14. The redshift–number relation for quasar subsamples

We next examine the quasar samples treated in the last section from the standpoint of the theoretical versus observed $N(< z)$ function. The observed and theoretical fractions, obtained by dividing respectively by the total number of quasars in the sample, or by the total volume of space out to the maximum redshift in the sample, are given in Figures 20–23. It is clear at a glance that the chronometric curve fits on the whole very well, but that the expansion-theoretic curve is in gross disagreement. This impression is fully confirmed by Kolmogorov–Smirnov tests, as indicated in Table 20. The indications given by the DeVeny heterogeneous list are fully supported by the more homogeneous samples.

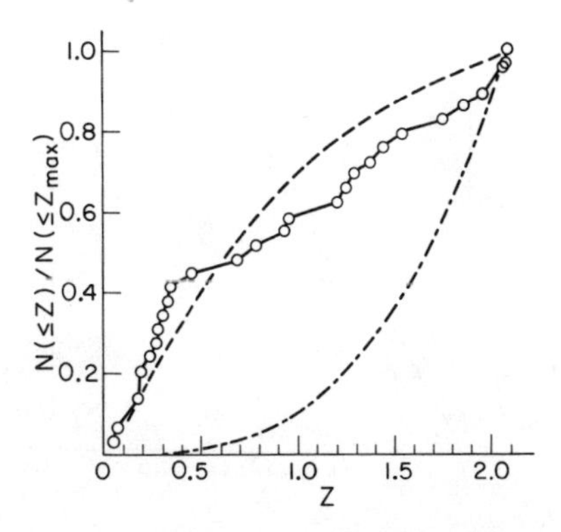

Figure 20 *The N–z relation for the quasar sample of Braccesi et al.* (1970).

Theoretical lines, assuming spatial and temporal homogeneity and approximate randomness of sample, are: —, chronometric theory; - - -, Hubble theory. Of the samples of quasars treated here, this sample of relatively faint quasars probably involves the maximal selection on luminosity and spectrum. Nevertheless it is in satisfactory statistical agreement with the chronometric prediction on the basis of a Kolmogorov–Smirnov test.

As earlier, it might be argued that due to luminosity selection, these samples are not adequately random, and that this circumstance is the origin of the apparent gross deviation from the expansion theory. However, in this event the agreement between observation and theory should improve substantially if the sample is cut off at a lower redshift. Such a cutoff diminishes the sample size and thereby the significance level of any given deviation, but it can otherwise not produce satisfactory agreement between the observed

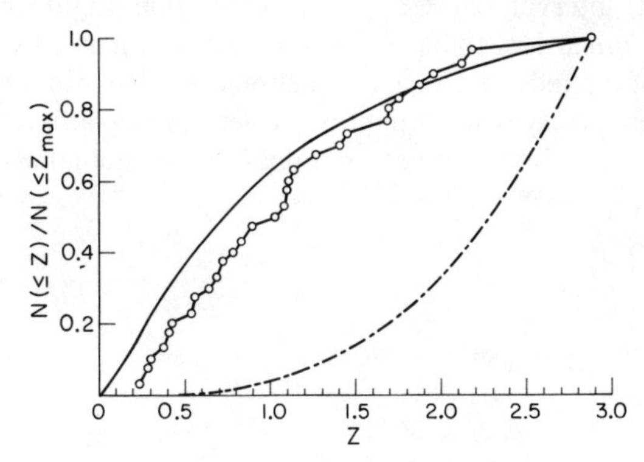

Figure 21 *The same as Figure 20 for the complete sample of* 4C *quasars given by Lynds and Wills* (1972).

○, individual quasar. The noticeable but not statistically significant apparent deficiency in the number of quasars at lower redshifts in this and the next sample may plausibly arise from the exclusion of quasarlike galaxies (notable Seyferts and N) which are found at these redshifts.

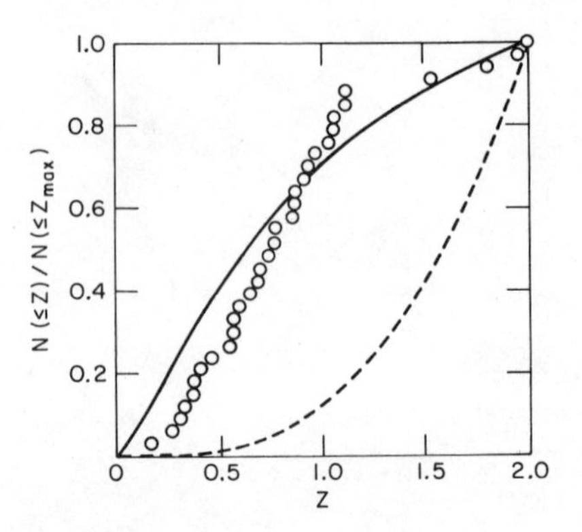

Figure 22 *The same as Figure 21 for the complete sample of 3C quasars given by Schmidt* (1968).

Compare the comment on Figure 21. Again the chronometric prediction for a random sample is in satisfactory statistical agreement with the observations.

$N(< z)$ and expansion-theoretic curves. For it is apparent that the slope of the observational curve is generally decreasing, in all cases, while the slope of the expansion-theoretic curve is materially increasing, at all redshifts. Adjustment by the scale factor involved in a cutoff at a lower redshift cannot change the sign of the second derivative of the $N(< z)$ curve, and so cannot eliminate this fundamental difference between the observations and the expansion theory.

In order to limit as much as possible extraneous sources of dispersion, to which deviations from the Hubble theory could conceivably be ascribed, two further tests were made. First, the samples were considered over a shorter redshift interval, $0.2 < z < 1$, in which one might anticipate some evolution, but much less than for the full redshift intervals of the samples. The deletion of the redshift range $z > 1$ should serve to diminish greatly any selection on luminosity which might be present in the samples. The deletion of the range $z < 0.2$ should serve to eliminate local anomalies. Nevertheless,

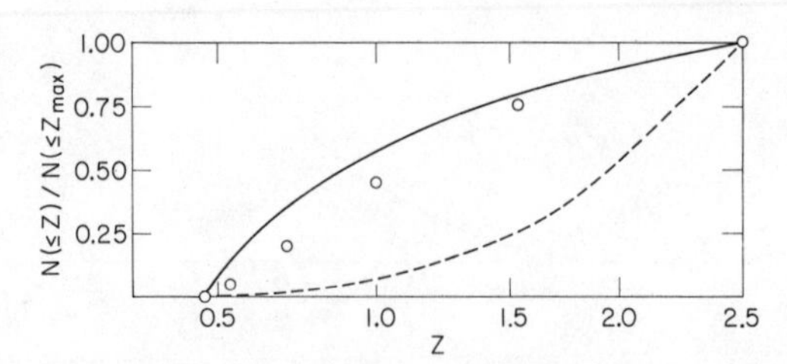

Figure 23 *The same as Figure 21, for the Schmidt adopted distribution* (1972a) *of optical quasars of approximate magnitude* 18.

Note: The plotted points represent summary data (observations on individual quasars not reported).

TABLE 20
Kolmogoroff–Smirnov test of the $N(z)$ relation for complete quasar samples

Sample	Number in sample	Chronometric D	Chronometric probability	Hubble D	Hubble probability
Schmidt 3C (1968)	33	0.19	0.16	0.58	$< 10^{-11}$
Schmidt optically selected[a] (1970)	19	0.24	0.21	0.49	$< 3 \times 10^{-4}$
Braccesi (1970)	27	0.22	0.14	0.58	$< 10^{-8}$
Lynds–Wills 4C (1972)	30	0.19	0.23	0.61	$< 10^{-9}$
Schmidt adopted redshift distribution of quasars of approximate magnitude 18[a] (1972a)	number N not given	0.14	> 0.05 if $N < 90$	0.50	$< 10^{-4}$ if $N > 20$

[a] Individual quasars were not listed, but only subtotals in specified redshift intervals. The D statistic used is the maximum over those z values for which data were given; and is therefore probably a slight underestimate of the true value.

as shown by Figure 24, the Hubble curve remains in gross disagreement with the observations, while the chronometric line fits very well, considering the limited sample sizes. The quantitative probabilities based on Kolmogorov–Smirnov tests are given in Table 21, in which, in addition, the DeVeny sample considered earlier and a sample of unspecified selection but substantial size given by Sandage (1972c) are included.

Second, the DeVeny sample was taken over the restricted redshift interval $0.25 \leqq z < 2.25$ as a means of removing local effects and minimizing possible confusions between quasars, N-galaxies, and Seyfert galaxies at the lower end, and of avoiding the apparently anomalous cutoff at the other end, which may reflect changes in the spectral functions of quasars at higher frequencies, or other relevant but largely unexplored effects. The results shown in Figure 25 are again in excellent agreement with the chronometric theory and in gross disagreement with the Hubble theory. Results for the Sandage sample over the complete redshift range of the sample are shown in Figure 26, and show agreement similar to that of the results in Figure 25. All available evidence, including the list of Burbridge and Burbridge (1969), indicates that all reasonably comprehensive or complete samples are likely to show the same behavior as the samples earlier treated (cf. Figure 33).

A still more conclusive acceptance of the chronometric and rejection of the expansion hypotheses (both on a nonevolutionary basis and the assumption of approximately uniform spatial distributions for quasars) can be obtained from the Schmidt V/V_m test, treated next.

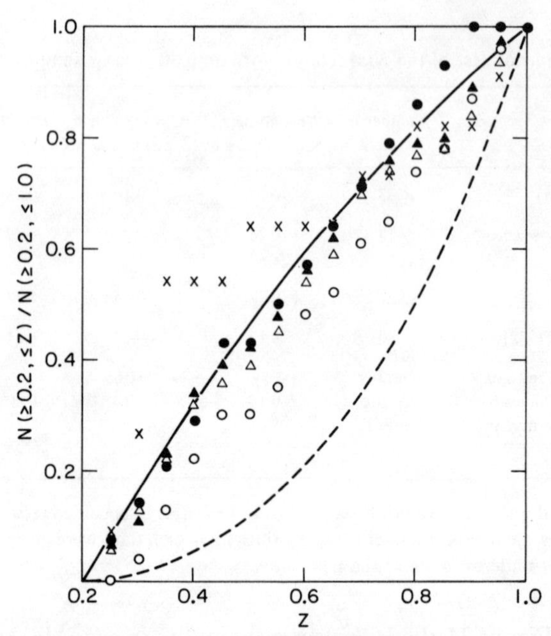

Figure 24 *The N–z relation in the redshift range* 0.2 < *z* < 1 *for quasar samples.* (a) ○, Schmidt 3C sample; (b) ●, Lynds-Wills 4C sample; (c) ▲, DeVeny list; (d) △, Sandage (1972b); (e) ×, Braccesi list. Otherwise on the same basis as Figures 20 and 21. The elimination of redshifts > 1 and the avoidance of possible local irregularities and classification difficulties by eliminating the region $z < 0.2$ do not materially improve the agreement of the expansion prediction with the observations. However, the chronometric prediction is in satisfactory agreement with the observations for all of the samples.

TABLE 21
The N–z relation for quasars in the range $0.2 < z < 1$ for diverse samples

		Probability of observed maximum deviation as given by Kolmogoroff–Smirnov test	
Sample	Sample size	Chronometric theory	Hubble theory
DeVeny	119	~ 0.5	$< 10^{-15}$
Sandage	77	~ 0.5	$< 5 \times 10^{-9}$
Braccesi	11	> 0.2	< 0.01
Lynds–Wills	14	> 0.4	~ 0.03
Schmidt	22	> 0.2	~ 0.08

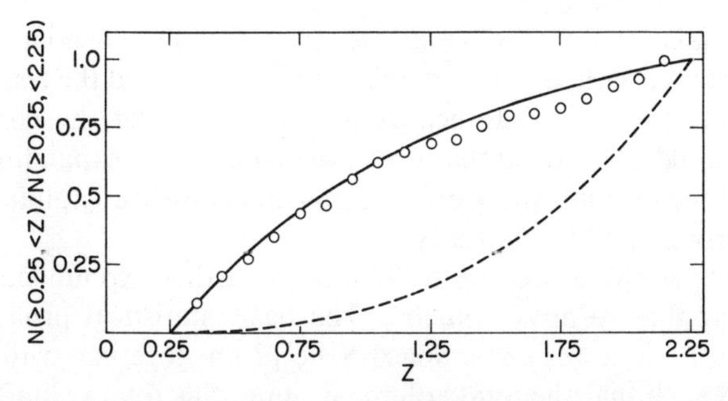

Figure 25 *The N–z relation for quasars in the redshift range* $0.25 < z < 2.25$.
Data: all quasars in DeVeny list with unquestioned redshifts and magnitudes. The cutoff above $z = 2.25$ used here corresponds to an observational one, and if removed would accentuate the discrepancy between the Hubble curve and the observations. A hypothetical extraordinarily broad luminosity function for quasars might serve to render the Hubble curve acceptable in relation to the observations, but would not explain the excellent agreement with the chronometric prediction. The deletion of quasars with $z < 0.25$ serves to remove from the comparison possible extraneous influences which cannot be resolved at this time. Compare the comment on Figure 24.

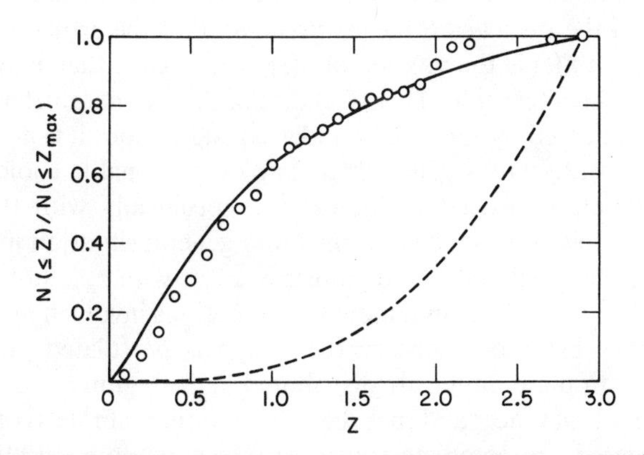

Figure 26 *The N–z relation for the quasar list of Sandage* (1972b).
The basis here is the same as earlier, except that the z values are limited to multiples of 0.1. Although no explicit selection criterion is given for the sample, it would appear on the chronometric hypothesis to be random in its redshift distribution.

15. The Schmidt V/V_m test for quasars in the chronometric theory

When samples of a specified type of luminous object are available which are complete down to specified limits of apparent luminosity, this provides a relatively universal and simple test for spatial uniformity. The availability of the Schmidt 3C quasar sample, the Lynds–Wills 4C quasar sample, and the Peterson galaxy sample—these are among the statistically most objective and substantial data on hand—indicates the importance of adapting the test to the chronometric theory. It is not difficult to do so, and the tests provide a significant measure of assurance as to the validity of the chronometric theory. It should be recalled that on a nonevolutionary expanding-universe theory, all three samples show quite strong and statistically quite significant deviations from spatial uniformity.

In principle, the procedure of Schmidt,† further expounded by Lynds and Wills, applies to any geometry. The basic statistical principle of the V/V_m test is as follows. Let a space S be given, together with a volume element in S such that the total volume is finite; this total volume may then be normalized to the value 1. This volume element may be entirely arbitrary apart from the requirement that individual points have zero volume. Let $\{S(t)\}$ be a one-parameter family of subsets of S, which are continuously increasing with S, and such that every point of S is contained in some $S(t)$, while no point is contained in all $S(t)$. For an object uniformly distributed in S, let $\bar{t}$ denote the least (or greatest lower bound) of the values of t for which the object is contained in $S(t)$. The volume of $S(\bar{t})$ is then a random variable V which takes on all values between 0 and 1. Furthermore, uniformity of the distribution of the object in S means precisely that the probability that the object will be in $S(t)$ is the volume of $S(t)$. This means that V is uniformly distributed in the interval [0, 1]. The choice of the one-parameter family $S(t)$ is in practice dictated by the theory under consideration; it is not mathematically unique, but there is generally a simplest reasonable choice.

The theoretical procedure for dealing specifically with the Schmidt V/V_m test will now be described, in the more general situation in which one considers only a fixed redshift region, $z < z_{max}$, with z_{max} not necessarily equal to ∞, but large. The fundamental assumptions involved in the analysis are then somewhat more conservative, for it is postulated only that the objects in question are uniformly distributed in the region $z < z_{max}$, and not necessarily in all of space; and that the observational sample(s) on which the analysis is based are complete (or constitute a random selection from a complete sample) only within the same region. Observationally, there is doubt as to the degree of accessibility of large redshift regions; as noted, e.g., by Burbridge (1971), the spectral function of quasars for large frequencies may fall off increasingly rapidly, making their observation more difficult (cf. in fact the last point in the observation of 4C 05.34 reported by Oke, 1970); spectroscopic selection may well be a factor in establishing redshifts and thus establishing that suspected quasars are indeed such (cf., e.g., Basu, 1973); intergalactic absorption, if present, would further limit the statistical validity of the inclusion of large redshift regions in the analysis.

† This appears to originate in part in work of P. Kafka (1967).

Consider, then, the region S of space in the chronometric theory in which $z < \tilde{z}$, where $\tilde{z}$ is a fixed arbitrary value, taken for relevance and simplicity to be > 1. For the most part, the regions $S(t)$ can be defined in the same way as in the expansion theory, as those out to a given redshift. There is, however, one case in which this is not possible. For an object in a complete sample at a redshift $z_1 > 1$, which is only slightly brighter than a limiting luminosity, and has a sufficiently flat spectrum, the region in which the object would be included in the sample is not the region of space out to a certain redshift, but consists rather of two disconnected pieces, of the form $z < z_1$ and $z_2 < z < \tilde{z}$. In this case, it would be incorrect to use the region below a given redshift. There is a natural choice of the one-parameter family of regions which is correct according to statistical theory, namely those of the form $z < t$ and $t' < z < \tilde{z}$, where $t < t'$ and the redshifts t and t' represent equal apparent luminosities for the spectral index in question. (Compare the earlier treatment of the chronotheoretic N–S relation.)

In the cases of the Lynds–Wills 4C and Schmidt 3C samples, most objects are sufficiently bright relative to the limiting magnitudes to be included in the sample wherever located in the redshift region $z < z_{max}$ on the basis of the chronometric theory. Of the remaining objects none actually involve the pair of disconnected regions just described, in the redshift region $z < 3$. In, e.g., the Schmidt sample only one object (3C 323.1, of spectral index 0.66) is radio-limited and its spectrum is too steep to lead to the disconnected regions just described, in the relevant redshift regions. Chronometrically, only two objects in the Schmidt sample, 3C 191 and 3C 9, are optically limited; in the Lynds–Wills sample, one object (4C 18.34) is radio-limited and one (4C 12.39) is optically limited. In all these cases, the spectra are too steep to lead to disconnected regions.

The results are shown in Figures 27 and 28. The horizontal axis is the V/V_m for the individual quasar; this is naturally theory-dependent, so there are two sets of points for the same observational datum. The value $z_{max} = 3$ has been used, expressing the possibility that completeness in the region $z > 3$ is best not assumed. This has substantially no effect on the expansion value of V/V_m, which have consequently been taken unchanged from the cited authors.

As was to be expected from the near independence of the apparent luminosity of quasars from their redshift on the chronometric theory, the corresponding V/V_m test gives results which differ by relatively little from the $N(z)$ comparison. In most cases the V_m for a quasar will be unity. The uniformity of the distribution of the V/V_m for quasars is then largely tantamount to the $N(< z)$ for quasars being proportional to the $V(z)$, and the Kolmogorov–Smirnov test for the uniformity of the V/V_m distribution is correspondingly related to the Kolmogorov–Smirnov test detailed earlier for the observed versus theoretical $N(< z)$ relation.

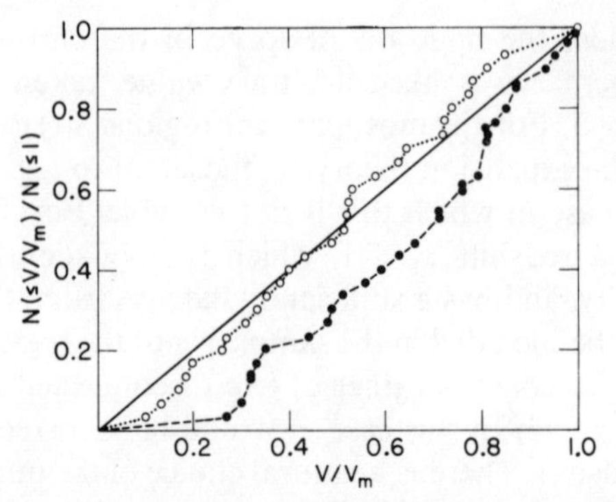

Figure 27 *The Schmidt V/V_m test for the 30 quasars in the complete sample of Lynds and Wills.*

○, chronometric analysis; ●, values given by Lynds and Wills for the Friedmann model with $q_0 = 1$; —, theoretical spatial uniformity. The chronometric values accept the hypothesis of spatial uniformity without any indication of luminosity and/or number evolution, unlike the expansion values.

For the expansion hypothesis this is not the case since the theoretical luminosity varies strongly with redshift in all redshift ranges. There is indeed a difference between the V/V_m test and the $N(< z)$ and also $N(m)$ test (the latter being discussed by Longair and Scheuer, 1970b, and by Lynds and Petrosian, 1972). However, as shown by Schmidt (1968) and also Lynds and Wills (1972) on the basis of this test, the expansion hypothesis is in poor

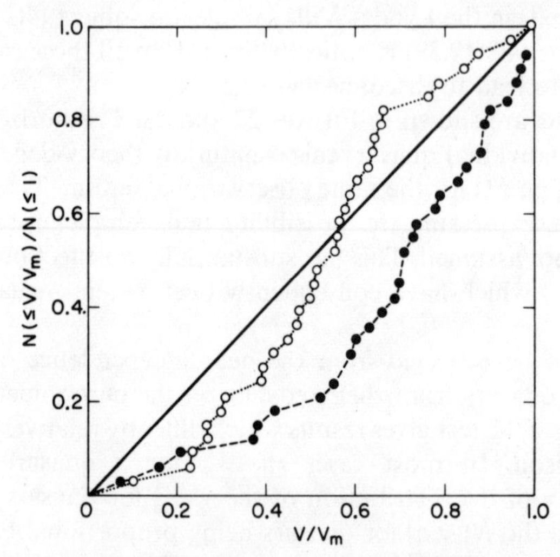

Figure 28 *The Schmidt V/V_m test for the 33 quasars in the complete 3C sample of Schmidt.* The basis here is the same as in Figure 27. The comment regarding that figure applies.

agreement with their data. Only with additional assumptions of luminosity and density evolution, as indicated by Schmidt (1972a,b,c), does the agreement become satisfactory, but this agreement is then virtually a matter of definition.

16. The angular diameter redshift relation for double radio sources

Given two theories regarding a variable y of the form:

$$y = f_j(x) + c \qquad (j = 1, 2),$$

where c is postulated to be an x-independent random variable of dispersion σ, it is evident that one can discriminate between the theories observationally with a moderate amount of data only if this dispersion σ is not too large compared to the average dispersion between the theories, i.e., the root mean square $f_1(x) - f_2(x)$ over the relevant range of x. In the case of the θ–z relation, this signifies that one can discriminate effectively between the relations

$$\theta_{\text{app}} \propto (1 + z)/z^{1/2} \qquad \text{(chronometric)},$$

$$\theta_{\text{app}} \propto (1 + z)^2/z \qquad \text{(standard cosmology)},$$

or variants thereof, only if the intrinsic dispersion in $\log \theta$ is not large relative to the average difference between the respective theoretical laws.

The intrinsic, model-independent dispersion in $\log \theta$ can be estimated from data which includes sufficiently many objects that many pairs at approximately equal redshift occur. Taking the largest and most thoroughly documented data, those compiled by Miley (1971), one finds 22 pairs of quasars which are double radio sources at approximately equal redshifts (the largest logarithm in the ratio of the pairs being 0.047, most being much less). If θ and θ' denote the angular diameters for such pairs, the quantity $[(2n)^{-1} \sum (\log \theta - \log \theta')^2]^{0.5}$, where n is the number of pairs, is a statistically consistent estimate of the dispersion in $\log \theta$, assuming that the distribution of θ_{int} is z-independent. This estimate is found to be 0.44. On the other hand, the root mean square of $\log(1 + z)/z^{1/2} - \log(1 + z)^2/z$ over the redshifts included in the sample is 0.03; and this would not be significantly altered by using a slightly different expansion-theoretic form, e.g., the Einstein–de Sitter angular diameter, or the Euclidean one. Thus the dispersion in the data is of the order of more than 10 times the theoretical difference to be probed, and no statistically significant comparison can be obtained. The Legg samples are smaller in size, and otherwise similar; hence they are likewise unable to discriminate between different theoretical θ–z relations of the type considered here.

The actual dispersions computed with the use of the Miley data are 0.36 and 0.37 respectively for the deviations of the observed $\log \theta$ from the expansion-theoretic and chronometric predictions, respectively. It was impossible for the two dispersions to differ by more than 0.03, so the closeness of these values was to be anticipated. It is somewhat unusual that the dispersions are less than the intrinsic dispersion, but the diminution in variance below the intrinsic level is not nearly at a significant level. The Miley data include 50 quasars which are double radio sources whose angular diameters are not indicated as questionable, and the dispersion of $\log \theta$ for these data is 0.79; thus the reduction of dispersion in $\log \theta$ itself via either theory is substantial.

For galaxies, the Legg data give dispersions of 0.56 and 0.52 respectively for the deviations of $\log \theta$ from the Hubble and chronometric theories respectively. Qualitatively the results are similar to those for the Miley data, and serve to confirm the conclusions just reached.

17. Observation versus theory for radio sources

There seems to be agreement among major surveys on two qualitative features of the N–S relation: (a) the elevation of the index $\beta = -\partial \log N/\partial \log S$ above the Hubble "Euclidean" value 1.5 (and a fortiori larger than attainable Friedmann values); (b) the decrease of N with increasing S, with $\beta \gtrsim 1$. In view of the uncertainty in the intrinsic luminosity function for radio sources, and the lack of published tabular data required for statistical analysis, it would be quite difficult to effect a Kolmogorov–Smirnov test of the data vis-à-vis the chronometric and expansion hypotheses, nor would any such test be conclusive at this time. It appears that little more can be said than that features (a) and (b) are difficult to reconcile with a nonevolutionary Friedmann cosmology, but are predicted by the chronometric theory, as earlier indicated.

The value of β may become infinite for a single luminosity class in the chronometric theory, but decreases rapidly as the breadth of the luminosity function increases. The values given in Figure 4 for a one-decade breadth in luminosity function agree reasonably well with the observations of Kellermann *et al.* (1971) and those of Pooley and Ryle (1968), when the latter are corrected for spectral index (cf. Kellermann *et al.*, 1971); the N/N_0 curve for the totality of radio sources should approximate the average of that in Figure 4 and the constant value 1, assuming an average spectral index ~ 0.6. Roughly this order of magnitude for the breadth, on the basis of the chronometric theory, is indicated by an analysis of the data presented by Schmidt

(1972b), Table 1, giving a list of 41 3CR sources complete in a given field up to a given limit and including redshifts for all but six sources. On the expansion theory, the intrinsic luminosity F_{rad} is given by the relation

$$\log F_{\text{rad}} = \log f_{\text{rad}} + 2 \log z + \text{const};$$

a computation of the dispersion of the log F_{rad} computed from this relation for the 29 galaxies having precise redshifts in the sample gives $\sigma = 1.02$; for all the objects having precise redshifts, consisting in addition of six quasars, the dispersion is much higher. On the chronometric theory, the dispersion is $\sigma = 0.55$ for the 29 galaxies of the 35 objects having precise redshifts, and somewhat greater if all 35 objects are included.

The relation between the dispersions on the two theories is consistent with that found earlier, and cannot be regarded as a coincidence. It indicates that bright radio sources are more nearly standard objects than had been thought, and suggests that observations down to fainter limits and of additional redshifts may yield quite discriminatory cosmological information. However, on the chronometric hypothesis there is presently no significant evidence whatever that radio sources have been evolving either in luminosity or space density. In addition to the cited data of Schmidt on radio luminosities, the following recent lists of redshifts versus radio luminosities are extant: 4C quasars (Lynds and Wills, 1972); 3C quasars (Schmidt, 1968); radio galaxies and quasars (Sandage, 1972c). The Lynds–Wills and Schmidt lists have explicitly designated completeness features; the Sandage lists are larger but the criteria for inclusion are not given explicitly. It is interesting that in all cases (which are not entirely independent, the quasar lists being overlapping), the chronometric dispersion in the luminosity–redshift relation is less than or approximately of the same size as the dispersion in apparent luminosity, while the dispersion from the expansion-theoretic line ($q_0 = 1$) is 50 to 100% greater. The specific values are given in the following table. Radio magnitudes are uncorrected and have been converted to the Pogson scale to facilitate comparison with the visual magnitudes. The apparent radio luminosities were not given in Sandage (1972c), but were reconstituted by J. F. Nicoll (unpublished course paper, MIT), in accordance with the equation $m_R = 5 \log z - 2.5 \log L_R$, where L_R denotes the absolute expansion luminosity tabulated by Sandage; see Table 22.

Thus the phenomenological superiority of the chronometric luminosity–redshift relation extends to radio luminosities. Concomitantly, the breadth of the radio luminosity function is highly model-dependent. From the chronometric standpoint, the breadth is quite moderate; indeed, substantially the full radio luminosity function for the quasars and galaxies may well be observationally accessible in the next few years.

TABLE 22

Data source	Number and nature of objects	Dispersions in: Apparent radio magnitude	Chronometric absolute radio magnitude	Expansion absolute radio magnitude
Lynds and Wills (1972)	30 quasars (4C list)	0.79	0.84	1.51
Schmidt (1968)	33 quasars (3C list)	0.75	0.73	1.27
Sandage (1972c)	68 radio galaxies	1.24	1.36	2.38
Sandage (1972c)	132 quasars	1.23	1.26	1.79

18. The Setti–Woltjer quasar classes

Setti and Woltjer (1973) have proposed that the m–z relation for sufficiently pure classes of quasars may be in rough agreement with the Hubble line. They have identified three classes, selected on their radio spectra, and found that the first class (those with steep spectra) show a definite trend with redshift, described as "a clear Hubble relation." The lack of such behavior for the relations of the other two classes is ascribed to a broad luminosity function.

It is interesting that the chronometric theory provides a considerably better fit for the m–z relation of all three quasar classes than does the Hubble line. In addition, the observational $N(< z)$ relations may be compared with those theoretically predicted on the assumption of spatial homogeneity of the quasar distribution. The chronometric prediction provides a very good fit, but the Friedmann model predictions for values of q_0 generally thought realistic are in gross disparity with the observations.

In particular, the important corollary to the Setti–Woltjer study that quasars are at essentially cosmological distances, and their redshifts are increasing functions of distance, is unaffected; but if the much better-fitting chronometric theory is correct, the distances are probably an order of magnitude less than those given by realistic Friedmann models, and $(\partial \log z/\partial \log r) = 2r/\sin r$, measuring r in units of R, i.e. z increases very rapidly with distance in much of the quasar redshift range.

a. *The m–z relations*

The dispersions in apparent magnitudes, and of the residuals of these magnitudes from the respective theoretical predictions are given in Table 23. All data satisfying the Setti–Woltjer criteria, and having unquestioned values of z, have been included in each sample. σ stands for the standard deviation of the indicated quantity; M represents an absolute magnitude, the subscripts c and e referring to the chronometric and expansion theories respectively, the latter being represented by the Friedmann model with $q_0 = 1$.

TABLE 23
The m–z relation for the Setti–Woltjer quasar classes

Sample	Size	$\sigma(m)$	$\sigma(M_c)$	Expected $\sigma(M_c)$	$\sigma(M_e)$	Expected $\sigma(M_e)$
Q_S ($\sim$ steep spectrum)	38	1.03	0.81	0.96	0.91	$0.72i$
Q_F ($\sim$ flat spectrum)	22	1.15	0.89	1.08	1.00	$0.69i$
Q_0 (radio quiet)	53	0.96	1.05	0.50	2.35	$2.23i$
Q_0, $z \leqq 0.5$	25	0.98	0.89	0.73	1.36	$1.16i$

In any theory of the form $m = f(z) + M$, where $f(z)$ is an analytically prescribed function of z, while M is an intrinsic magnitude, it is to be expected on an elementary statistical basis that $\sigma(m)^2 = \sigma(f)^2 + \sigma(M)^2$, or $\sigma(M) \sim [\sigma(m)^2 - \sigma(f)^2]^{1/2}$. The latter quantity has been computed and entered in the table as "expected dispersion." Pure imaginary values signify that the dispersion in apparent magnitudes is less than would be expected on the hypothesis that the m–z relation has the form indicated; the absolute value of the expected dispersion is then an indication of the extent to which the hypothesis deviates from expectation, on the basis of the sample in question.

The subsample of the optical quasars for which $z \leqq 0.5$ has been included because it is large enough to be meaningful, and in order to minimize possible selection on luminosity, which is in all likelihood greatest for this particular type of quasar. It serves also to test, and actually to confirm, the a priori reasonable idea that the excess of $\sigma(M_c)$ over $\sigma(m)$ for the total optical sample is a consequence of the effective cutoff in apparent magnitudes for the optical quasars.

For each sample, the $\sigma(M_c)$ is substantially less than the $\sigma(M_e)$; for all samples except one, the $\sigma(M_c)$ is less than the $\sigma(m)$, and is in reasonably good agreement with the expected $\sigma(M_c)$. The $\sigma(M_e)$ is less than the $\sigma(m)$ for the first two samples, but much greater than it for the optical sample; and in all cases the expected $\sigma(M_e)$ is quite different from the actual $\sigma(M_e)$.

The Q_S and Q_F samples are the purer subgroups identified by Setti and Woltjer. For Q_S this means $\alpha \geqq 0.7$ and a suitable double radio structure. The Q_F have $\alpha < 0.6$ and are of relatively small angular diameter in a sense specified by Setti and Woltjer (designated P in their preprint).

b. The $N(< z)$ relations

On the assumption of spatial homogeneity, the $N(< z)$ for each type should vary approximately with the volume out to redshift z. The extent to which this is the case is shown by Figure 29. The chronometric prediction is in visibly good agreement with the observations; the Friedmann model with $q_0 = 1$, and the Hubble model, are respectively in poor and very poor agreement with observation. In all cases the redshift range has been limited to $0 < z < 2$, for uniformity and to minimize probable selection effects at higher redshifts; the sample sizes are thereby slightly reduced.

A quantitative measure of the deviations of the $N(< z)$ counts from theory is afforded by the Kolmogorov–Smirnov statistic D. This is the maxi-

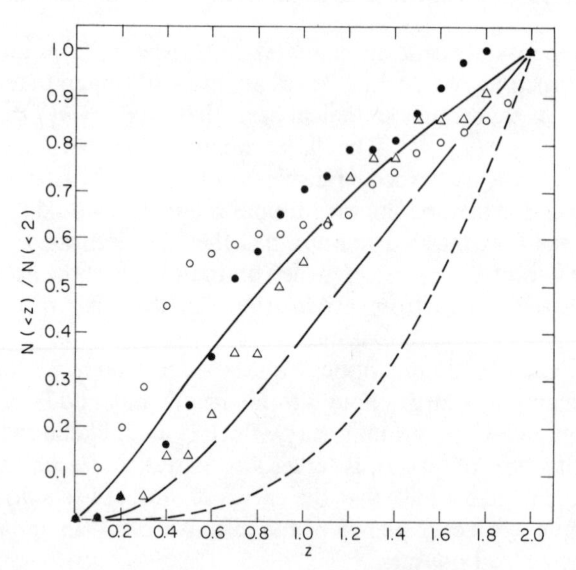

Figure 29 *The redshift distributions of the Setti–Woltjer quasar classes in comparison with theoretical predictions.*

—, chronometric theory; – –, Friedmann model with $q_0 = 1$; - - -, Hubble theory. ●, Q_S; △, Q_F; ○, Q_0. Despite the observational selection which may be present, the chronometric prediction is statistically acceptable on the basis of Kolmogorov–Smirnov tests for all classes.

TABLE 24
Radial spatial homogeneity of the Setti–Woltjer quasars as measured by Smirnov deviations and probabilities

Sample	Size	D_c	$P(D_c)$	D_e	$P(D_e)$
Q_S, $z < 2$	37	0.11	$\gtrsim 0.5$	0.31	1.4×10^{-3}
Q_F, $z < 2$	20	0.12	$\gtrsim 0.5$	0.22	0.29
Q_O, $z < 2$	46	0.25	7.6×10^{-3}	0.43	6.0×10^{-8}
Q_O, $z < 0.5$	25	0.07	$\gtrsim 0.5$	0.34	6.2×10^{-3}

mum absolute difference between the fractions observed and predicted up to redshift z. It is given together with the corresponding probabilities in Table 24, the D_e referring to the Friedmann model with $q_0 = 1$.

Schmidt V/V_m tests, based on an assumption of effective randomness of the samples in the totalities of quasars of each type with $z < 2$ and reasonable prescribed limiting luminosities, can be expected to give similar although less definitive results, in view of the analytic similarities between the tests, and experience with other quasar groups.

The chronometric predictions fit extremely well, if the reasonable assumption is made that the optical sample suffers quite materially from luminosity selection when taken out to redshift 2; this assumption is indicated by the good fit for the subsample out to redshift 0.5. As in a number of earlier samples, the strong degree of spatial homogeneity the results indicate, on the chronometric hypothesis, naturally raises a question, if the hypothesis is accepted, as to the possible existence of gravitational or other dynamical effects which tend to enhance spatial homogeneity, or at least to maintain it. On the other hand, the Friedmann model predictions are rejected at conservative probability levels, indicating once again that the spatial distribution of quasars can be reconciled with the expanding-universe theory only on the assumption of very strong evolution.

The Q_F data are accepted by both the chronometric and expanding-universe theories, but they are less appropriate than the other data as a check on theoretical predictions. This is due to the relatively small Q_F sample size, and to the possible model-dependence of the type, whose criterion involves a restriction on angular diameter in fixed angular measure, of which the metric implications are theory and z-dependent.

c. *Discussion*

The results are similar to those earlier presented for less homogeneous quasar samples. It is also possible to analyze in a similar way the θ–z relations, but again as earlier the large dispersion in log θ in the vicinity of fixed z precludes any strongly indicative comparison between the chronometric and expansion theories on this basis.

The hypothesis that the quasar m–z and θ–z relations are purely stochastic is a specific form of the "local" hypothesis which can be rejected on a definitive statistical basis by tests of trend in the m–z and θ–z relations of the Q_S, confirming earlier more informal analyses by Miley (1971) and Setti and Woltjer. In particular, the Spearman rank-correlation test gives a probability $\sim 10^{-5}$ of obtaining a value of the Spearman coefficient as large as that observed, between m and z, if these variates are stochastically independent. This test is entirely independent of any assumption as to the distribution of m or z; it is also model-independent. The same test can be conducted with the Q_S sample replaced by the DeVeny sample of 158 quasars earlier described; it is interesting that the results are of the same order of definitiveness, due to the lesser homogeneity of the sample. The Q_S thus appear as one of the purest and statistically useful classes of quasars yet identified.

19. Other observational considerations

Finally we turn to a number of phenomena, or anomalies from the expansion-theoretic standpoint, which do not primarily involve statistical testing.

a. The energy requirements of quasars vis-à-vis bright galaxies

Many authors have cited the unprecedented energy requirements of quasars on the expansion-theoretic hypothesis as the major anomaly associated with them. Many new, largely quite speculative, hypotheses have been proposed to explain the mechanism of the energy output. Most of these hypotheses are of a partial or qualitative nature which renders them immune from direct statistical testing.

There is no difficulty whatever regarding the energy production of quasars on the chronometric hypothesis. A straightforward analysis indicates that they are within ~ 1 mag of nearby bright galaxies, brightest cluster galaxies, Seyfert or Seyfert-like Markarian galaxies, and N-galaxies; cf. Table 25. Possible intergalactic absorption is too small to be important in the present connection, as shown by the consistently small chronometric dispersion for the bright quasar m–z relation. More concretely, the relative intrinsic luminosities of galaxies and quasars are compatible with the hypothesis that quasars are the nuclei of certain relatively luminous galaxies, whose outer portions are invisible at larger redshifts. There is other evidence

TABLE 25

Luminosities of bright extragalactic objects: averages ± standard errors

Data source	Sample and size	Absolute chronometric magnitude	Apparent magnitude	Absolute expansion ($q_0 = 1$) magnitude
de Vaucouleurs tape (1964)	15 galaxies of redshift nearest 250 km sec^{-1}	18.72 ± 1.21	11.02 ± 1.18	26.43 ± 1.26
de Vaucouleurs	15 galaxies of redshift nearest 500 km sec^{-1}	18.55 ± 1.09	11.60 ± 1.06	25.51 ± 1.13
de Vaucouleurs	15 galaxies of redshift nearest 1000 km sec^{-1}	17.86 ± 0.87	11.66 ± 0.86	24.05 ± 0.87
de Vaucouleurs	15 galaxies of redshift nearest 2000 km sec^{-1}	18.04 ± 0.60	12.59 ± 0.60	23.47 ± 0.60
de Vaucouleurs	15 galaxies of redshift nearest 4000 km sec^{-1}	18.20 ± 0.80	13.49 ± 0.80	22.87 ± 0.81
Peterson[a] (1970a)	44 bright cluster galaxies, complete to 15^m	17.47 ± 0.48	13.71 ± 0.82	21.16 ± 0.33
Sandage[a] (1972b)	41 bright cluster galaxies	17.78 ± 1.37	14.75 ± 2.50	20.58 ± 0.31
Sargent (1972)	24 Seyfert-like Markarian galaxies	19.01 ± 0.77	15.43 ± 0.89	
Burbridge and Burbridge (1967)	All 74 galaxies with data in list	18.27 ± 0.91	17.32 ± 1.09	17.83 ± 1.29
Lynds and Wills (1972)	All 30 quasars with data in list	18.69 ± 0.89	17.84 ± 0.99	18.03 ± 1.32
Sandage (1972b)	All 109 radio-noisy uasars with data in list	18.20 ± 0.89	17.26 ± 1.03	17.75 ± 1.30
DeVeny *et al.* (1971)	All 157 with unexceptionable data	18.42 ± 0.93	17.39 ± 1.02	18.11 ± 1.59
Wampler *et al.* (1973)	OQ 172 (redshift 3.53)	17.77	17.5	14.76
Sandage (1972b)	15 radio-quiet quasars	19.20 ± 1.08	17.74 ± 0.98	19.56 ± 2.52

[a] Magnitudes uncorrected for difference between expansion-theoretic apertures reported and chronometrically correct apertures.

for this hypothesis, and no significant evidence from the chronometric standpoint against it; thus from this standpoint, quasars are not significantly more exotic than galactic nuclei in which apparently violent activity is present. It is the qualitative nature of these activities, rather than the average energy output of the galaxies in question, which may suggest an unconventional explanation of their energy source.

More specifically, the best-fitting chronometric redshift–magnitude curve to the quasars listed by DeVeny *et al.* takes the form (assuming spectral index ~ 1)

$$m = 2.5 \log z - 2.5 \log(1 + z) + 18.4.$$

We define the "chronometric intrinsic magnitude" of an object of magnitude m and redshift z as the number c such that

$$m = 2.5 \log z - 2.5 \log(1 + z) + c.$$

This number is not necessarily independent of the location of the object due to possible deviation of the spectral index from 1, possible intergalactic absorption, etc. The advantage of this measure of intrinsic luminosity is that it is independent of the assumed value of the Hubble parameter; it should be borne in mind that only relative intrinsic luminosities are meaningful in the present sense. Table 25 details the relative intrinsic luminosities of various types of quasars and galaxies.

b. *The sharp decrease in the number of quasars in the range* 2.2–2.9 *and apparent near cutoff beyond* 2.9

This phenomenon is too well known observationally to require description. A variety of explanations has been proposed from the expansion-theoretic standpoint, all involving a greater or lesser degree of ad hoc assumption, and the introduction of one or more new parameters. There seems no reason to doubt that spectroscopic selection effects play a partial role; also, it becomes progressively more difficult to establish the larger redshifts, and it is reasonable to anticipate that intergalactic absorption or obscuration will increase with distance, and so with redshift in either the chronometric or the expansion theory. While these effects may largely explain the near cutoff above redshift 2.9, the order of magnitude of the attrition in the region 2.2–2.9 of demonstrated accessibility is more difficult to understand on the expansion-theoretic hypothesis.

To make a simple order-of-magnitude estimate, suppose to begin with that the quasars are approximately of the same intrinsic luminosity, and are uniformly distributed in space. The ratio $N(2.2 < z < 2.9)/N(z < 2.2)$ for quasars is then 1.29 on the Hubble theory, $\geqq 0.33$ for a Friedmann model with $|q_0| \leqq 1$, and 0.09 on the chronometric theory. In the DeVeny list there are 200 quasars in the range $z < 2.2$ and 5 in the range $2.2 < z < 2.9$. The expansion-theoretic expected number in the latter range is 66–258, on the basis of the 200 observed in the former range. This is a discrepancy too great to be eliminated by reasonable modification of the luminosity function, or the spatial distribution, or to be explicable by spectroscopic selection of any known type. It is also perhaps beyond the need for formal statistical analysis; but on the hypothesis of approximately random selection of 205 quasars from all those of redshift < 2.9, and those indicated regarding the luminosity function and spatial distribution, the equivalent Gaussian variate represented by the standard approximation to the Bernoulli distribution is 15.5 on the Hubble theory, and $\geqq 9.9$ for a Friedmann model with $0 \leqq q_0 \leqq 1$, corresponding to probabilities $< 10^{-53}$ and 10^{-23}, respectively. Clearly drastic supplementary hypotheses are required to explain the cutoff on the expansion theory, and the proposed explanations are of this nature.

In the chronometric theory, the expected number of quasars is 17.6. The discrepancy from the observed value of 5 is 12.6, a number of quasars which might well arise from random fluctuations combined with possible sharp drops in the spectral function of quasars shortward of the Lyman α line, tendencies toward spectroscopic selection, and intergalactic absorption and/or obscuration. The formal Gaussian variate is here 3.1, just beyond the conventional significance level, the corresponding probability being 0.002. The discovery of just five additional quasars in the range $2.2 < z < 2.9$ would render the deviation insignificant by conventional standards (probability > 0.05). As indicated by the subtlety of the identification of 4C 05.34 as a quasar by Lynds and Wills (1972), confirmation that suspected quasars in this range are indeed such is relatively difficult to supply, and is in part anticipated on the chronometric theory according to which there should be relatively few other quasars at nearby redshifts available for comparison purposes. The difficulty of supplying this confirmation may well be partly responsible for the apparent slight deficiency in the number of quasars observed in the range in question on the chronometric hypothesis.

It may be noted finally that the preceding section was written prior to the discoveries of two quasars at redshifts ~ 3.5. These discoveries and the attendant circumstances regarding colors are consistent with and indeed support the foregoing, but they exacerbate the discrepancy between a priori expansion-theoretic indications and actual observation.

c. *Superlight velocities*

It is evident that these apparent velocities are highly sensitive to the estimate of distance, and so to the redshift–distance relation. With the chronometric relation, all published apparent superluminal velocities are reduced to well below the velocity of light. Even if H were as low as 40 km sec^{-1} Mpc^{-1} at Virgo, instead of the larger value taken in the bulk of this paper, the largest apparent superluminal velocity, that of 3C 279 would be reduced to less than c. With $H = 80$ at 10 Mpc, its velocity would appear as $(0.57 \pm 0.17)c$. On the basis of the conventional expansion theory, its apparent velocity is $\sim (10 \pm 3)c$; see Whitney *et al.* (1971). While other explanations of superluminal velocities have been given (cf. Cavaliere *et al.*, 1971), the present one appears to be the scientifically most economical, in requiring no assumptions beyond the fundamental one of the chronometric model itself.

d. *The relative absence of quasar identification for radio sources at faint magnitudes*

Bolton (1969), Braccesi *et al.* (1970), and Fanti *et al.* (1973) have independently found that quasar identifications are relatively rare for radio sources on plates which are sufficiently sensitive to record objects of visual magnitude in the vicinity of ~ 20.5. Galaxy identifications have been made in this way, and on the expansion-theoretic hypothesis, there is no apparent reason why significant numbers of quasars should not appear on such plates. Such relative lack of quasars is, however, predicted by the chronometric theory, augmented by the earlier validated hypothesis that quasars form approximately a single luminosity class on which is superimposed a Gaussian variate of dispersion $\lesssim 1$ mag. Compare Figure 17, in which the small and decreasing slope of the chronometric $N(m)$ curve at $m = 19.5$ is indicative of the cutoff.

e. Apparent distance-dependence of and variation in the Hubble parameter

The persistent anomaly in the determination of the Hubble parameter by different observations is largely removed by the chronometric theory, as detailed in Section 8 of this chapter.

f. The cosmic microwave background radiation

As earlier noted, virtually any strictly temporally homogeneous theory of the cosmos will predict a blackbody background radiation, as the equilibrium photon gas formed by the free radiation in the universe. Conservation of energy and maximization of the entropy dictate the blackbody form, in the presence of ergodicity (cf., e.g., Mayer 1968). The latter is implied by any significant degree of overall stochastic perturbation, which in the real Cosmos arises from the evolution and motions of galaxies, scattering, absorption, and reemission by intergalactic matter, possible gravitational deflections, etc.

Turning now to comparison with observation, it is natural to inquire how well the observed relative energy of the background radiation and starlight of $\sim 10^3$ conforms to the theoretical analysis earlier given, which yielded an upper bound and putative order-of-magnitude estimate for the background energy. This was based on the assumption that emission and absorption by bright galaxies are major factors in the establishment of the background equilibrium radiation. Defining "bright" as of magnitude ≤ 13 at a redshift $cz = 2000$ km sec^{-1}, the catalog of de Vaucouleurs (1964) indicates ~ 250 bright spiral galaxies in this redshift region, indicating $\mu \sim 1.4 \times 10^5$ (using now units with $R = c = 1$). The angular diameters given in the catalog correspond to a metric diameter of ~ 10 kpc on the chronometric hypothesis, assuming $z = 0.005$ at a distance of 15 Mpc (which incidentally limits the age of pristine radiation to $\lesssim 10^9$ ly). The value $r = 5$ kpc $= 1.2 \times 10^{-5}$ then is indicated if the absorption in the Galaxy, $\sim 0.3 \csc b$, is reasonably typical. The resulting theoretical prediction for the ratio of the energy of the microwave background to that of starlight is 1.7×10^5, an excess over observation corresponding to a black-body temperature of ~ 2 times that observed. This seems fully comparable in precision to the accuracy of the prediction based on the "big-bang" hypothesis in view of the hypothetical parameters, such as the entropy density of the original universe, involved in the latter prediction. Faint galaxies and/or unknown forms of intergalactic matter could account for the discrepancy, as could also in part the probable underestimation of r resulting from the existence of significant dust exterior to the observed optical diameters of the galaxies.

g. Cosmic time scales

The universal and relativistic scales are nonlinearly related, indeed an infinite relativistic time corresponds to a finite universal time. The ages derived by radioactive dating and other microscopic considerations are relativistic, on the chronometric assumption that local elementary particle interactions are effectively observed by a relativistic clock. This position represents the minimal departure from conventional theoretical practice, and follows from the unicity of this clock implied by Lorentz and scale covariance.

The coincidence of the orders of magnitude of the apparent ages of the earth, sun, galaxy, etc., is understandable on this basis in the following way. At a fixed point taken as the origin in space, the relation between the universal and relativistic time is $\tan t = x_0/(1 - x_0^2/4)$, in units of R, earlier estimated as ~ 106 Mpc. All observed ages correspond to values of t in the range $-\pi < t < 0$, since $x_0 = 0$ when $t = 0$ and $x_0 = -\infty$ when $t = -\pi$. If the universal age of discrete objects in the universe is uniformly distributed in this range (to which attention may be confined, since older discrete objects would not be observable as such, light from them going through an infinite redshift), the corresponding relativistic age will follow a nonuniform but calculable distribution in the range $-\infty < x_0 < 0$.

Specifically, the uniform distribution law for t in the range $-\pi/2 < t < \pi/2$ is found (by calculus) to correspond to the Cauchy distribution law whose element over the range $-\infty < x_0 < \infty$ is $(2\pi)^{-1}(1 + \frac{1}{4}x_0^2)^{-1}\, dx_0$. Restricting consideration to the observable range $x_0 < 0$, the integral $\pi^{-1} \int_0^\infty x(1 + \frac{1}{4}x^2)^{-1}\, dx$ expressing the expected age of a random object of any specified type is divergent. It follows that the arithmetic means of the ages of a sample of such objects (e.g., galaxies) should fluctuate widely. However, the percentile points should converge to the corresponding percentile points in the overall population. The 50, 95, 97.5, and 99 percentile ages (i.e., the ages such that the given percentage has a lesser age) are 2, 25.4, 50.9, and 127.3, in units of the time for light to travel a distance equal to the radius of the universe. With $R = 106$ Mpc as earlier, this means that half of the galaxies should have ages in the range $\sim 0.7 \times 10^9$ to 10^{10} yr. This value of R corresponds to the value $H = 100$ at 15 Mpc; with the larger value $R \sim 150$ Mpc corresponding to the parameter $H = 50$ at 15 Mpc the corresponding range is simply $\sim 10^9$–10^{10} yr. The result obtained here appears to be in good agreement with the limited number of independent estimates or bounds on galaxy ages, and serves to explain within the chronometric framework the coincidence of the order of magnitude of the apparently older astronomical objects.

h. *Holmberg's systematic effect in galaxy clusters*

One of the striking features of extragalactic redshifts which appears at variance with theoretical anticipation is the "extremely high internal redshift dispersion found for clusters of nebulae," in the words of Holmberg (1961), who first noted and analyzed this effect, most notably by a precise and detailed treatment of the Virgo cluster. Holmberg shows that relatively conventional explanations such as "short lifetimes or tremendous gas contents" are unrealistic, and that the results found would be implied by a systematic effect of magnitudes on redshifts. While such an effect is physically different from a nonlinear redshift–distance relation, it is mathematically closely related, in that a suitable such relation will lead to effects such as those analyzed by Holmberg.

Indeed, the chronometric prediction is in satisfactory agreement with the data listed by Holmberg. His principal data are for the Virgo cluster, and are given as his Table 1. The dispersion in intrinsic velocities varies as the dispersion in absolute magnitude. For the expansion theory, this is quite large, indeed significantly greater than that in apparent magnitude; but for the chronometric absolute magnitudes, the dispersion is only slightly greater than that in apparent magnitude. The quantitative results vary with the particular subsample involved, but the qualitative results do not. The respective dispersions are: (a) in apparent magnitude, (b) in deviations from the Hubble line, (c) in chronometric absolute magnitude: (1) for all 84 nonblueshifted galaxies which are listed, (a) 1.15, (b) 1.69, (c) 1.29; (2) for all 46 nonblueshifted So or E galaxies (which as indicated by Holmberg are of particular interest in relation to the question of estimation of the mass of the cluster), (a) 1.30, (b) 1.71, (c) 1.39.

Another apparent effect of a generally similar nature has been detailed for the Coma cluster by Tifft (1972). As in the case of the Virgo cluster, analysis of the data precisely as listed, but on the basis of the chronometric rather than expansion theory, leads to an acceptably small apparent cluster redshift dispersion. Again the qualitative results are independent of the particular morphological type. The respective dispersions are (1) for all 70 galaxies with data, (a) 0.57, (b) 0.67, (c) 0.59; (2) for all 28 ellipticals, (a) 0.56, (b) 0.72, (c) 0.61; (c) for the 42 nonellipticals, (a) 0.51, (b) 0.63, and (c) 0.54.

V

Discussion

1. General conclusions

Substantially all potentially relevant published systematic data on extragalactic objects have played a part in the foregoing parallel tests of the chronometric and expansion hypotheses. The satisfactory, and for the most part strikingly good, agreement of the chronometric predictions with the raw observations is in clear contrast with the only rarely really good agreement of the direct expansion-theoretic predictions with corrected observations. When consideration is confined to samples that are complete out to specified limiting magnitudes, the comparison is still more one-sided in favor of the chronometric theory; there is no such sample, either of galaxies or quasars, which is at all well fitted by the unembellished expansion theory. Furthermore there is no present indication that more refined studies are at all likely to alter the basic fact that the chronometric theory, with essentially no free parameter, provides a much better overall fit to extragalactic data than do straightforward general relativistic models with the free parameters q_0 and Λ. Indeed the trend of recent work has largely been in the opposite direction.

This is not to say that the expansion theory has been disproved. Its generally idealistic, nonoperational nature is readily compounded by the introduction of ad hoc mechanisms—superclustering for nearby galaxies, number-luminosity evolution for quasars, and other features reminiscent of epicycles—which may serve to render moot its apparent disagreement with the observations. But saving the theory in this way largely elminates its predictive power, and by virtue of the effective increase in the number of adjustable parameters, renders it scientifically highly uneconomical in comparison with the chronometric theory.

In any event, the remarkable degree of observational confirmation of the chronometric theory naturally raises the question of whether there is any sound scientific reason not to employ this hypothesis in theoretical astrophysics in preference to the expansion hypotheses. The expansion theory has for many years enjoyed the status of a preferred theory, with its concomitant influence on both the direction of observational research and its quantitative results. Inevitably questionable observations tend to be resolved in conformity with an established theory, while conversely observation in apparent conformity with the theory tend to be regarded as relatively unexceptionable. This general feature of experimental science is particularly important in an area in which facilities have been extremely limited, in which observations are not readily repeated by independent observers, and in which there is inherently little control or capacity for more intensive examination of the objects under study. The highly limited telescope time suitable for extragalactic work, and the intrinsic restriction to the information obtainable from the observation of their electromagnetic emission, imply that astrophysics falls into this category to the *n*th degree. The inability of the expansion theory to make useful fundamental observable predictions, despite its dominance over the past 40 years, is in striking contrast with the capacity of the chronometric theory to predict accurately a broad variety of theory-independent relations derived from observations published prior to its existence. This suggests that the chronometric theory is, at the least, likely to be relatively useful as a framework for the organization and study of observations on extragalactic objects.

Figure 30 is an illustration of the coherence which the use of the chronometric theory can introduce into the study of the nature of different types of extragalactic objects. The generally highly satisfactory fit of the theory to the samples on which Figure 30 is based, as well as a number of other samples, is shown in Figures 31 and 32. No substantial or otherwise cogent published samples of galaxies or quasars are unrepresented in these figures, except that due to Colla *et al.* (1975), which was published too late to be included in these figures, but which as earlier noted has the same qualitative implications as the large or otherwise statistically cogent galaxy samples.

It has to be admitted that the square redshift–distance law predicted for low-redshift objects is in striking variance with a generation of instruction in cosmological astrophysics, and at first glance appears to be contradicted by Sandage's observations on brightest cluster galaxies. While referring to

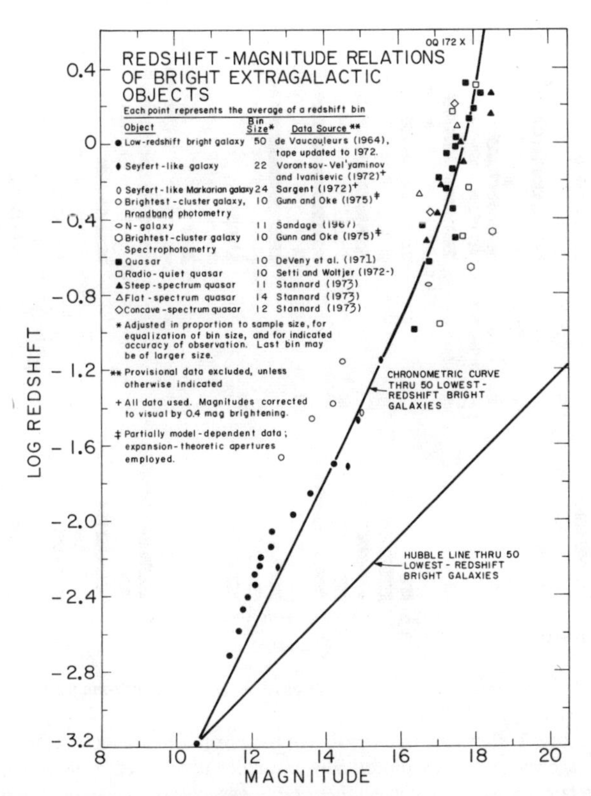

Figure 30 *Redshift–magnitude relations of bright extragalactic objects.*
Sources: de Vaucouleurs and de Vaucouleurs (1964), Gunn and Oke (1975), DeVeny *et al.* (1971), Sargent (1972), Vorontsov-Vel'yaminov and Ivanisevic (1974), Sandage (1967), Setti and Woltjer (1973), and Stannard (1973). Thanks are due the authors of the last two cited sources for communicating the data on which their graphs and other reduced results were based.

Chapter IV for a detailed analysis of the latter point, a decent regard for the natural prejudices and conservative proclivities of those brought up scientifically on the expanding universe seems to require an attempt to explain how so fundamentally misleading an apparent observational result could become so firmly imbedded in astrophysical thinking. There are sociological and biographical matters here which while probably quite interesting are beyond the scope of this book, and of the author's competence, and require additional information which is not readily available. But some nontrivial illumination is derivable from material in the scientific literature.

The boldness of Hubble's first paper (1929), which was in all probability influenced in part by theoretical considerations, as emphasized by Hetherington (1971), was one factor. The small sample of galaxies studied in this

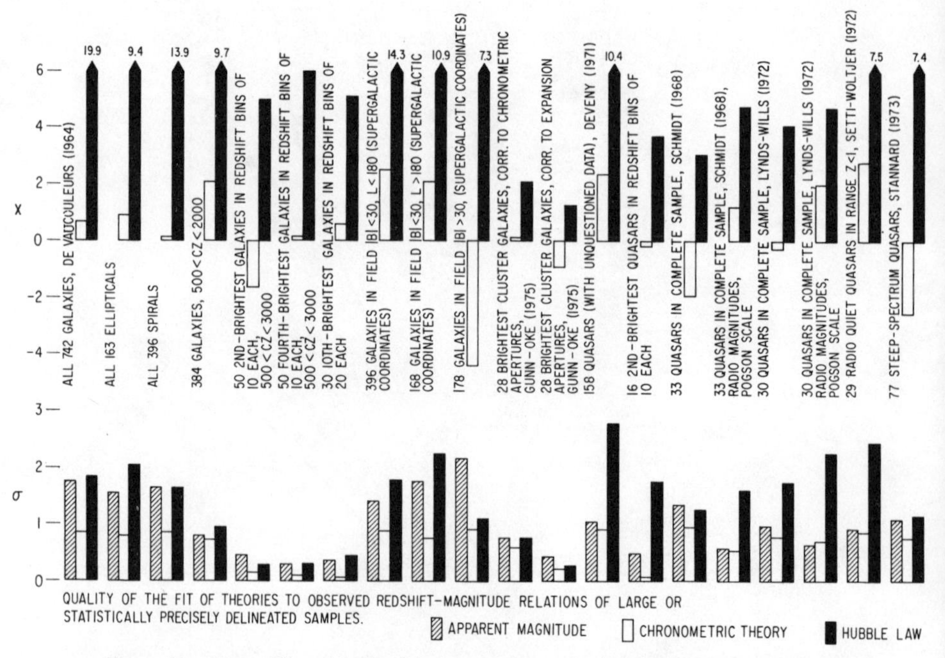

Figure 31 *The quality of the fit of theories to observed redshift-magnitude relations of large or statistically precisely delineated samples.*

The quantity X shown is the negative of the normalized reduction in variance brought about by the theory and is asymptotically normally distributed with zero mean and unit variance, for a fair sample and a correct theory. The variances σ^2 are in the apparent magnitude and in the residuals from the chronometric and Hubble law predictions. (In each case, the zero points of the predictions are adjusted to the mean luminosity of each sample.) The tendency toward order-of-magnitude equality of the variances in apparent magnitude and in residuals from the Hubble law for galaxy samples (or others at $z \lesssim 0.5$) is predicted by the chronometric theory, as is the larger variance in these residuals for larger-redshift samples. The subsamples specified by supergalactic coordinates were defined ex post facto (by G. de Vaucouleurs) in accordance with a theory indicating different redshift-distance exponents in the respective regions of the sky; the X values are correspondingly equivocal, but favor the square law over the linear one even in the region $|B| > 30$ hypothetically maximally favorable to the linear law. See also Tables 11 and 12. Data sources, in addition to those for Figure 30: Lynds and Wills (1972) and Schmidt (1968).

paper had an observed m–z relation which is distinctly better fitted by a square redshift–distance law than by a linear one. Nevertheless Hubble described the law as "roughly linear," on the explicit basis of uncertain and surely rough estimates of distance to only 10 of the galaxies; and the implicit basis of the prior theoretical prediction of a roughly linear law by a suitable development of general relativity. Later the estimates of distances to galaxies was refined by a study of the "brightest stars" in low-redshift galaxies, leading

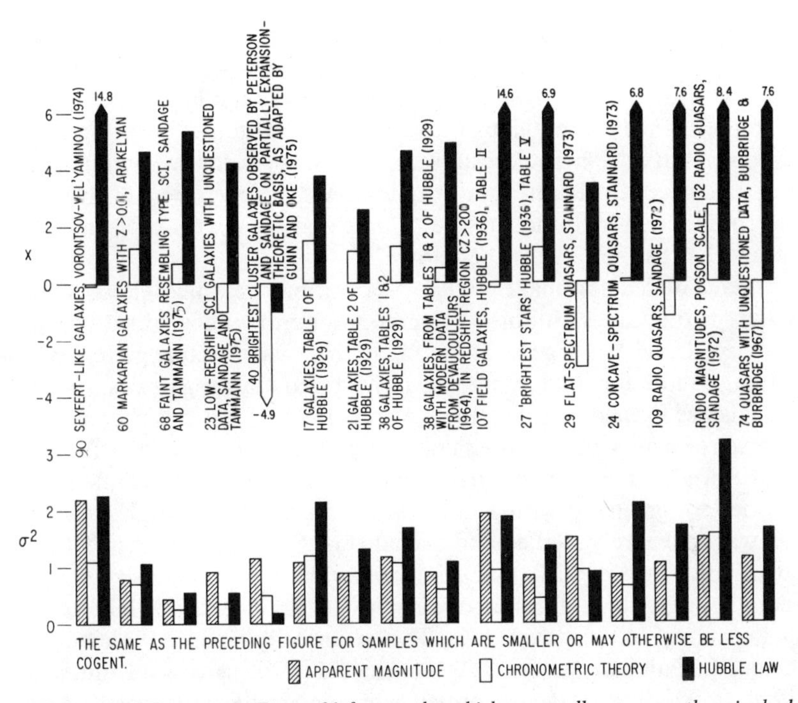

Figure 32 *The same as Figure 31 for samples which are smaller or may otherwise be less cogent than those in Figure 31.*

Data sources, in addition to those cited in Figures 30 and 31: Arakelyan *et al.* (1972), Sandage and Tammann (1975), Hubble (1929, 1936), and Burbridge and Burbridge (1967).

to Hubble's conclusion that these were suitable distance indicators. But the observed m–z relation of these "stars" is again in distinctly better agreement with a square redshift–distance law than with a linear one; and again, there is no indication in Hubble's work of this fact, or that he was at all cognizant of it.

In collaboration with Humason, Hubble (1931, 1936b) made additional observations, more than 100 field galaxies being included in their sample published in 1936. Again, this sample is in much better agreement with a square than a linear law, but mention or cognizance of this is not in evidence. Of course, the lack of agreement with a linear law could always be ascribed to an extreme breadth of the luminosity function for the galaxies observed; but such conceivable agreement with a linear law is very different from a positive indication for it. Moreover, if one attempts to suppress or limit the effect of the breadth of the luminosity function by utilizing only the brightest galaxies in bins ordered by redshift, the results are equally favorable to the square law vis-à-vis the linear law. But in these papers appeared for the first time the class of galaxies which were later to form the foundation for the linear redshift–distance law which had been proposed by Hubble in 1929. It consisted of bright cluster galaxies, and was a relatively much smaller sample than those previously considered by Hubble; however, its m–z relation was in extremely close agreement with the linear law prediction. The papers included no definite indication of how the ten clusters in this sample were chosen, and with the passage of time and the deaths of the authors, it appears that this may never be known.

It was difficult to make similar observations on other telescopes. Over the next two decades Hubble's original program was developed by Humason, Mayall, and Sandage, culminating in their classic paper giving redshifts and magnitudes for a large number of field and cluster galaxies. Again, the field galaxies formed a large sample whose m–z relation was in considerably better agreement with the prediction of a square rather than a linear law; and the brightest cluster galaxies formed a relatively small sample in somewhat better agreement with a linear law. This sample of bright cluster galaxies was intensively studied and extended by Sandage over the next decade and a half, during which time preliminary results were reported by him, generally in graphical form, and fairly widely accepted as definite proof for the linear redshift–distance law. The superiority, indeed unicity, of the 200-in. Mt. Palomar facility, and the decades of intensive study initiated by Hubble, made difficult the performance of comparable work elsewhere. With the publication of numerical observational results by Sandage in 1972, the basic case for the linear law seemed to be finally documented by the observed m–z relation for the sample of 41 brightest cluster galaxies which he treated. But as detailed earlier, analysis of the $N(z)$ relation and the apparent Hubble core radii for galaxies in the sample naturally raises a question of apparent selection effects, which cannot be dispelled on the basis of published information. The very satisfactory agreement of the phenomenological m–z relation for other types of galaxies with the prediction from a square law naturally reinforces the apparent anomaly of the Sandage sample.

The main moral is perhaps the importance of taking effective cognizance of the distinction between model-building on the one hand, and hypothesis-testing on the other. Another is the need for the exploration of foundational observable relations by several independent groups of observers. Finally, the great difficulty and expense of the observations do not supersede the need for utilizing the best available robust statistical analysis, but on the contrary, strongly enhance the marginal utility of the extraction of all relevant information from the samples. This need is closely related to that for statistically unexceptionable sampling procedures, involving notably the designation of explicit objective criteria for the inclusion or exclusion of objects from the sample.

In work published since the original manuscript of this book was completed, Rust (1974) has given a precise study of data for 36 supernovae, including particularly time-delay estimates. Here it can only be commented that the objects treated by Rust appear to be among the best "standard candles" available, but that the sample is too small for definitive analysis, and that the time-delay effect is theoretically less clear-cut than those treated in Chapter IV (cf. the discussion of observed versus theoretical times in Chapter III). The peak magnitude–redshift relation for all non-blueshifted objects (with or without removal of unrepresentatively large redshift objects) is in quite satisfactory agreement with the chronometric prediction, but poor agreement with the expansion prediction. The $N(<z)$ relation in the redshift region $cz < 1000$, which is unlikely to be greatly affected by an observational magnitude cutoff, is also in closer agreement with the chronometric prediction, the observational estimate of $\partial \log N/\partial \log z$ at $z = 0$ being fairly close to the chronometric value, $\frac{3}{2}$.

2. Theoretical aspects

Although the present theory in principle alters only physical kinematics, and so is vastly more limited than any complete dynamical theory, the alteration is of a fundamental character, which may in consequence cause some reflection by specialists. It remains to reconsider, if necessary, a variety of developments in astrophysical theory in terms of the chronometric theory. Most fundamentally, the questions arise: (a) what is the relation to the theory of Friedmann models; more broadly, how does it relate to general relativity, or to gravitation as a purely physical process? (b) what observable consequences, if any, does it have as regards elementary particle phenomena? A brief discussion of these and some related questions follows.

a. Slow expansion?

In principle, there is no difficulty in combining the chronometric redshift theory with some degree of expansion in accordance with a closed Friedmann model. As long as the rate of expansion is kept sufficiently low, the excellent agreement of the chronometric theory with observation is sufficient to ensure a statistically acceptable fit of the combined theories in standard cosmological tests. Moreover, the basic features of the Friedmann theory and its correlation with cosmology, apart from the redshift itself, would be retained.

Such a mixed theory cannot be excluded on a purely statistical basis, and would permit conventional ideas concerning the evolution and age of galaxies to persist in the combined theory without essential change.

However, the remarkable quality of the fit of the theory to observation is impaired noticeably as the rate of expansion increases. This is particularly the case for some of the key nonstatistical matters: the problem of the energy output of quasars, the apparent near cutoff in quasars circa $z = 2.5$, and the apparent existence of superlight (or near superlight) velocities. In addition, it would generally diminish the economy and predictive power of the theory. It would therefore seem more interesting and promising to seek to reexamine conventional ideas on the age, evolution, and colors of galaxies, in the light of the pure chronometric theory, than to develop a combined theory at this time. There is no apparent reason to anticipate any greater difficulties in so doing than exist already at present.

As noted by Segal (1972), and discussed in Chapter III, the chronometric theory can be regarded as defining (and is largely defined by) a virtual motion of the canonical local Lorentz frame at each point of the Cosmos with respect to the same frame at any other point. This canonical frame is that tangential to the universal (chronometric) frame at the point in question, and has natural units specified by setting $\hbar = c = R$. Thus a virtual Doppler redshift is implicit in the chronometric theory. However, the virtual point of view has no physical advantages, but only the theoretical one of possibly facilitating the correlation with the formalism of general relativity. The canonical local Lorentz frames at different points differ in scale as well as by a conventional Lorentz transformation. This variation in scale can be removed only by making the radius of the universe time-dependent, since the distance scale is fixed chronometrically by setting $R = 1$. Having made this change in scale, one has a pure Doppler relation between the correspondingly rescaled local Lorentz frames, and a formally expanding universe. The velocity of this initial recession varies approximately as the square of the distance for small distances, but for larger distances is not the Doppler velocity in formal correspondence with the chronometric redshift.

b. Gravitation

In mathematical respects, there is no significant difference between the chronometric and the Minkowski models from the standpoint of general relativity as a local theory of gravitation. The chronometric model is conformally flat, indeed the physically relevant local correspondence between local chronometric space–time and Minkowski space–time is given in explicit analytic form in Chapter III.

The increasing but still not yet totally definitive validation of general relativity as a local theory of gravitation does not imply its physical applicability in the form of the theory of Friedmann or similar models. As a local theory of gravitation, it is one in a hierarchy of local dynamical theories, none taking precedence over the others; as a cosmological theory, it has a more fundamental and different status from the others. Being inherently lacking in temporal homogeneity, the conservation of energy in our earlier group-theoretic sense is violated, and indeed there does not yet exist a broadly accepted and fully viable definition of energy in classical (unquantized) general relativity. The time itself is defined in terms of the formalism, rather than observable physical processes. In these respects it differs greatly from elementary particle and quantum field theories, in which there are formally well-defined positive definite energies, and in which the time may be characterized uniquely, apart from choice of scale, by the constraint of temporal invariance, and observed directly in terms of a theoretically precise frequency standard.

It would seem distinctly metaphysical to extrapolate the mathematical formalism of general relativity from a theory valid on the galactic scale as one of a hierarchy of theories of different interactions, to a theory on which the dynamics of the entire universe must be based, and to which the clocks of elementary particle processes must conform. Probably still less justified physically is the application of general relativistic hydrodynamics to extragalactic questions such as the mass density and the stability of the entire Cosmos. The approximation of the distribution of galaxies by a fluid is quite uncontrolled and open-ended; at best, conclusions drawn in this way are merely suggestive. The astrophysically fundamental fact that much, if not most, of the mass of the universe is in the form of the discrete bound states called galaxies, is completely lost sight of in the process of this approximation, and may represent a more crucial physical point of departure than the study of overall mass density of the Cosmos.

Admittedly, as a nondynamical theory, the chronometric model is incapable of predicting the average density of the universe. But this separation between kinematics and dynamics is quite possibly the way it should be. The circumstance that there is basically no such separation in global general relativity, while philosophically striking and unique, can be regarded as a major source of the ambiguities in the elucidation of its precise physical meaning. The clear-cut separation between kinematics and dynamics in elementary particle theory has made for empirical lucidity of the theory, and has on the whole been very satisfying. Indeed, a strong current trend in general relativity has been toward its recasting in terms analogous to those employed in the theory of elementary particles and their associated quantum fields. Work by Faddeev (1971) on the correlation of general relativistic and quantum concepts, leading in particular to a possible appropriate notion of energy, represents fundamental progress in this direction. The chronometric model could serve equally well with Minkowski space in Faddeev's work; its use in place of Minkowski space would actually lead to simplifications, in that the delicate question of the appropriate boundary conditions at infinity in space is superseded by the closure of space.

Likewise adaptable to the chronometric framework is the important foundational work of Lichnerowicz (1961) on the quantization of general relativity.

As pointed out to us by C. C. Lin, the mass density given by the standard closed Friedmann model with fixed radius is, with the radius of the universe given by the chronometric theory of the order of 10^{-27} gm cm^{-3}. This is quite high, but perhaps not unacceptably so, particularly in the light of comparable difficulties with missing mass in conventional theory. It remains to be explored to what extent observational estimates of the mass density of the universe may be affected by employment of the chronometric rather than the expansion model as the theoretical substructure.

In a more theoretical vein, it is interesting to note possibilities for correlating the chronometric model with general relativistic local gravitational theory through the scalar field provided by the presently unspecified scale of the conformally flat metric involved in the model together with the vector that vanishes in the special relativistic limit, as in Section 9 of Chapter III. In this connection, mention should be made of Weyl's conformally oriented theory (1921) and of his projective tensor, having the feature that it vanishes if and only if the metric is conformally flat. It is interesting that in the very natural form of the theory presented by Veblen (1933) there intervene both a scalar and a vector field, such as are provided locally by the chronometric theory in the $R \to \infty$ limit. Moreover, the mere modification of a Minkowskian metric by the introduction of these two fields in the indicated fashion is sufficient to imply all the observational consequences of general relativity (cf., e.g., Hawking and Ellis, 1973). Finally, the frequently-conjectured elimination of singularities in general relativity by the introduction of quantum considerations seems closer in that the strict form of the theory, for finite R, involves the five hermitian operators whose approximately scalar form for large R gives rise to the indicated scalar and vector fields.

In summary, the chronometric theory, although based on a physically entirely different redshift mechanism from expansion or gravitation, is in both observational and mathematical respects otherwise compatible with much of general relativity, including all of its local ($\sim$ galactic) features.

c. *Elementary particles*

As indicated in Chapter III, the fundamental local dynamical variables of the chronometric theory, energy, momenta, etc., differ from those of special relativistic field theory by terms of order R^{-1}, or less, where R is the radius of the universe in conventional laboratory units. These dynamical variables are here regarded as generators of local symmetries. The difference is therefore too small to affect observationally any known elementary particle processes, apart from possible selection rules and classification features, assuming the state spaces to be the same in both cases. For particles of zero mass, the wave functions are locally essentially unchanged, but for massive particles the chronotheoretic structure remains to be developed. The existence of mass is not compatible with the transformation of elementary particles under the full conformal group; current ideas of broken symmetry and the like indicate, however, that a group of related massive elementary particles may well arise from restriction of this group (more precisely, of its universal covering group) to an appropriate mass-conserving subgroup.

The most obvious choice for this subgroup is that leaving fixed an observer's infinity and local distance scale; this is the conventional inhomogeneous Lorentz group. The square of the mass is then represented by the image under the relevant representation of the D'Alembertian $\square$. When combined with conventional representations for internal symmetries or quantum numbers, it would leave the theory of elementary particles and their local interactions basically unchanged. A physically more natural choice for the subgroup is however $O(2, 3)$. This also facilitates mass splitting, which is forbidden for the Lorentz subgroup by the O'Raifeartaigh theorem. The role of $\square$ would be taken over by the Casimir operator for $O(2, 3)$, which differs essentially from $\square$ only by terms of order R^{-1}.

d. *The mechanism of energy production in galactic nuclei and quasars*

Obviously any hypothesis regarding this matter can be validated only in a quite indirect fashion. From the chronometric standpoint there is, however, an extremely simple and natural supposition regarding the nature of the mechanism: it consists basically of the transformation of the excess of the unienergy $i^{-1}(\partial/\partial\tau)$ over the special relativistic energy $i^{-1}(\partial/\partial t)$, i.e., of the new form of energy corresponding to the difference between the two times involved in the theory (earlier shown to be positive) into elementary particle processes. As earlier noted, the excess unienergy appears with redshifting, and is then diffused in space in a fashion which causes no observable local particle production. The amount of energy involved is, however, quite substantial, and over very large distances and times should be responsible for significant interactions. The formation, dissolution, and early intensive development of galaxies would appear on the one hand to involve interactions of this nature, and on the other to be the most likely known physical process not clearly explicable by elementary particle combined with gravitational interactions.

If the conversion of the unienergy excess over the special relativistic energy into local particle processes is indeed a significant feature of galactic development, it should be one of the main mechanisms by means of which the unienergy excess energy density in the universe is kept stationary, as it would be natural to assume it is in the chronometric theory. White dwarfs and other elderly, metal-rich objects may provide fuel which can be ignited with sufficient unienergy excess to yield elementary particles and hydrogen. Such "burning" of moribund objects in galaxies could provide very large amounts of energy in quite small regions, and plausibly take place on a short time scale, as in the case of supernovae, giving rise to variability such as is observed in active galactic nuclei. At the same time it would serve to maintain the population of moribund objects at an approximately stationary level.

Highly speculative as these considerations are, they are perhaps less so than those treating the origin and early dynamics of the universe, which involve much larger regions of space and much longer reaches of time. The group-theoretic nature of the chronometric hypothesis implies a variety of conservation laws which are relatively stringent, physically meaningful, and to a considerable extent serve to define the theory. It is therefore operationally much more subject to definitive validation than is the "big-bang" theory and similar hypotheses.

In any event, because of the temporal homogeneity of the universal cosmos, there must be processes that convert the superrelativistic into the relativistic energy. This is effected in part by free propagation, which while leaving the total energy unchanged alters its partition between the delocalized superrelativistic energy and the microscopically observable relativistic energy. However, the energetics of the microwave background indicates that other processes must be more important.

Consider, for example, what may be the basic cycle involving the bulk of matter and radiation in the universe: gas + microwave background + dust + local radiation + old stars → galaxies → gas + microwave background + dust + local radiation + old stars. The local radiation density over regions of the order of 10 parseconds is naturally subject to large random fluctuations in the course of cosmic time, especially if enhanced by cosmic turbulence along the lines of von Weizsäcker (1951). These fluctuations should eventually reach the flash point required to "ignite" the other ingredients of a generic galaxy core. Of course, this does not exclude other possible mechanisms, such as collisions involving cores of old galaxies or old stars within the cores, which collisions necessarily take place because of the infinitude of time and the finiteness of space. A particularly interesting feature of the chronometric outlook for the evolution of galaxies is the availability of ample energy from the superrelativistic radiation for making up the mass loss from galaxies with active nuclei. This mechanism would both explain the otherwise persistently puzzling large mass loss, of the order of one solar mass per year (cf. Oort, 1974), and provide for additional large-scale conversion of the superrelativistic radiation into conventional forms. In particular, it would lower the estimated cosmic background radiation temperature from the correct order-of-magnitude but slightly high figure earlier derived.

e. ***Intergalactic matter and the microwave background***

The precise equilibrium attained by radiation in space, following possibly many complete circuits in space, will naturally depend on the extent and character of intergalactic matter which may be present. The existence of such matter proposed by Holmberg (1958) and indicated by some later studies (Takase, 1972, among others) is difficult to substantiate directly. A small rate of extinction, of the order of $\lesssim 10^{-6}$ mag/kpc if $H \sim 80$ at 10 Mpc is consistent with the redshift–magnitude relations for quasars and galaxies, in the chronometric theory. Because of the smaller size of the chronometric than the expansion-theoretic universe, relatively little dust would be required to produce this rate of extinction, but on the other hand relatively little would be necessary to play a significant role in the possible thermalization of intergalactic radiation. The very general analysis by Purcell (1969) (based on the Kramers–Kronig relation) indicates a density of matter $> 10^{-33}$ g cm^{-3}, assuming transmission characteristics not grossly dissimilar from those of interstellar dust and the indicated extinction rate, but there are no other known restrictions on the dust. Compare in this connection, the mechanism for production of the background radiation proposed by Layzer and Hively (1973).

A detailed analysis of how the spectrum of emitted radiation is transformed depends also on the absorption and emission characteristics of galaxies and intergalactic matter, as well as on having globally more precise wave functions for the emitted radiation than are afforded by simple plane wave. The chronometric theory predicts quite directly the existence of very nearly isotropic and highly energetic blackbody radiation diffused homogeneously throughout the universe. The temperature of this radiation is a dynamical quantity which the theory can only correlate with other dynamical quantities. This is effectively the case also with the primeval fireball concept. Indeed, the latter development involves more parameters than the chronometric prediction, and these parameters must be quite specially chosen in order to lead to an observationally correct prediction (cf. the careful account by Weinberg, 1972). On the whole, the observed cosmic background radiation is predicted by the chronometric theory in at least as definitive a fashion as the primeval fireball complex of hypotheses.

f. *The Friedmann model with* $q_0 = \infty$

To take q_0 as ∞ would exacerbate the missing mass problem in general relativistic cosmology, and otherwise appear at first glance not to be observationally sustainable. It has however been observed by J. F. Nicoll that the value $q_0 = \infty$ is in many ways the best-fitting of the Friedmann models to the general run of cosmological data, on galaxies and quasars. Its redshift–distance law is identical with that of the chronometric theory, and it thereby gives an equally good account of the $N(< z)$ relations for galaxies and quasars. Its m–z and N–S relations are however different by virtue of the "number effect" due to the recession. This results in a distinctly poorer fit to the quasar m–z relation than the chronometric relation. In the case of the N–S relation, there is the same difficulty as in other Friedmann models, that without evolution the values of the index $-\partial \log N/\partial \log S$ fall below the Euclidean value 1.5, in disparity with the radio source observations earlier described. From a purely theoretical viewpoint, the $q_0 = \infty$ case is one of the most interesting Friedmann models by virtue of its exceptional symmetry.

3. Further observational work

In view of the historical observational basis for the expansion theory in the data on low-redshift galaxies, the question arises of the existence of statistically rigorously appropriate data which are actually favorable to, and not merely perhaps marginally consistent with, this model. Its relation to the cosmic microwave background, coincidence of order of magnitudes of time scales, and the apparent helium abundance are dependent on stringent dynamical assumptions supplementary to the expansion hypothesis itself. Consequently, these relations only indicate the possibility, and not an objective or definitive likelihood, that the model is correct. The only clear possibility for the vindication of the predictive cogency of the expansion hypothesis appears to lie in systematic galaxy observations in randomized fields out to a limiting magnitude of ~ 15. But it is evident that what is to be anticipated on the basis of the present work is rather a *reductio ad absurdum* of the model.

In any event, the observations on each galaxy should include magnitudes over a range of apertures, in order to form an appropriate basis for testing several theories. The sample should either be complete in a designated field to specific limits, or constitute a randomized subsample from such a complete list. Radio observations might additionally be quite useful in this connection. Although there is no special reason to doubt that the results would be generally similar to those earlier obtained, such observations would then lead to more precise estimates of the radius of the universe, the number density and size of bright galaxies, and other important cosmological parameters.

The fact that increased accuracy and numbers of observations, over the years, has not appreciably altered the phenomenological situation as regards the m–z–N relations of galaxies, except strongly to confirm earlier indications appears equally true of quasars. The m–z and N–z relations of the ~ 70 quasars known approximately eight years ago are similar to those for the $\gtrsim 200$ quasars known today. For example, for the 74 quasars with unquestioned data listed by Burbridge and Burbridge (1967), the respective dispersions in apparent magnitude, deviation from the Hubble line, and in chronometric absolute magnitude are: (a) 1.09, (b) 1.29, (c) 0.92, which is qualitatively similar to the results from the later DeVeny and other samples, in the reduction and increase in dispersion associated, respectively, with the chronometric and expansion theories. As was to be expected, the average magnitudes are fainter for the later sample, but only by slight amounts: (a)

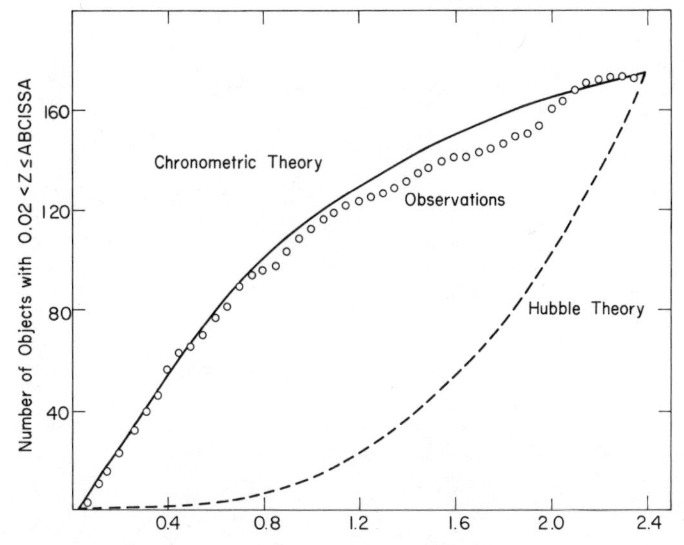

Figure 33 *The redshift distribution of the quasarlike objects with redshifts published as of* 1969.

All quasars, N and Seyfert galaxies listed by Burbridge and Burbridge (1969) for the redshift range $0.02 < z < 2.40$ were included. The redshift distribution is similar to that for the later list published by DeVeny *et al.* (1971) and the earlier list of Burbridge (1967), except that the inclusion of N and Seyfert galaxies removes the apparent deficiency in the number of quasars at redshifts $\leqq 0.3$. The m–z relations are also similar, and there is no apparent reason to expect that additional quasar observations over the next few years will materially affect their overall m–z–N relation.

0.07, (b) 0.32, (c) 0.15. The same is true of the $N(z)$ relation, which is qualitatively similar on the basis of today's observations to that of several years ago. Figure 33 shows in fact the excellent fit of the direct chronometric prediction to the redshift distribution of quasarlike objects listed by Burbridge and Burbridge (1969). Here "quasarlike" means that the object is listed either as a QSO, a Seyfert, or an N galaxy. As earlier indicated, inclusion of the latter removes an apparent if statistically not significant deficiency of quasars at redshifts < 0.3 from the chronometric outlook. Whether these objects are included or not, the observational relation is generally similar to that shown in Figure 25, based on the DeVeny sample. From the expansion-theoretic standpoint, quasars are murky, variable objects, among the least likely to provide an observational foundation for the theory. But larger samples of quasars which are complete out to fainter limits, or random subsamples of such, could lend additional confirmation to the chronometric theory; and in any event would help to clarify the nature of quasars, even if, as is to be anticipated, their basic redshift–luminosity–number relation is not substantially altered.

References

Abell, G. O. (1972). *In* "External Galaxies and Quasi-Stellar Objects" (Symp. No. 44), p. 341. International Astronomical Union, Reidel, Dordrecht, The Netherlands.
Abell, G. O., and Mihalas, D. M. (1966). *Astronom. J.* **71**, 635.
Alexandrov, A. D., and Ovchinnikova, V. V. (1953). *Vest. Leningrad gos. Univ.* **11**, 95.
Alexandrov, A. D. (1967). *Canad. J. Math.* **19**, 1119.
Arakelyan, M. A., Dibai, E. A., and Espisov, V. F. (1972). *Astrofizika* **8**, 33.
Bahcall, J. N., and Hills, R. E. (1973). *Astrophys. J.* **179**, 699.
Basu, D. (1973). *Nature* **241**, 159.
Baum, W. A. (1972). *In* "External Galaxies and Quasi-Stellar Objects" (Symp. No. 44), p. 393. International Astronomical Union, Reidel, Dordrecht, The Netherlands.
Berman, S. J. "Wave equations with finite velocity of propagation." Doctoral dissertation, MIT. Published: *Trans. Amer. Math. Soc.* **188** (1974), 127–148.
Bolton, J. G. (1969). *Astronom. J.* **74**, 131.
Braccesi, A., Ficarra, A., Formiggini, L., Gandolfi, E., Lari, C., Padrielli, L., Tomasi, P., Fanti, C., and Fanti, L. (1970b). *Astronom. and Astrophys.* **6**, 268.
Braccesi, A., Formiggini, L., and Gandolfi, E. (1970a). *Astronom. and Astrophys.* **5**, 264.
Burbridge, E. M. (1967). *Annual Rev. Astronom. and Astrophys.* **5**, 399.
Burbridge, E. M. (1971). *In* "Nuclei of Galaxies." *Vatican Conf. 1970* p. 121.
Burbridge, E. M., and Burbridge, G. (1967). *Nature* **222**, 739.
Burbridge, E. M., and Burbridge, G. (1967). Quasi-Stellar Objects. Freeman, San Francisco.
Busemann, H., (1967). *Dissertationes Math. (Rozprawy Mat.)* **53**, 52 pp.
Carruthers, P. (1971). *Physics Reports* **1C**, No. 1.
Carswell, R. F., and Strittmatter, P. A. (1973). *Nature* **242**, 394.
Cavaliere, A., Morrison, P., Sartori, L. (1971). *Science* **173**, 525.
Choquet-Bruhat, Y. (1971). *General Relativity and Gravitation* **2**, 1.
Colla, G., Fanti, C., Gioia, I., Lari, C., Lequeux, J., Lucas, R., and Ulrich, M.-H. (1975). *Astron. Astrophys.* **38**, 209.
DeVeny, J. B., Osborn, W. H., and Janes, K. (1971) *Publ. Astronom. Soc. Pacific* **83**, 611.

Ellis, G. F. R., and Hawking, S. W. (1973). "The Large-Scale Structure of Space-Time." Cambridge Univ. Press, London and New York.
Faddeev, L. D. (1971). *Actes, Congr. Internat. Math. 1970* **3**, 35.
Fanti, R., Formiggini, L., Lari, C., Parielli, L., Katgert-Merkelijn, J. K., Katgert, P. (1973). *Astronom. and Astrophys.* **23**, 161.
Fish, R. A. 1964, *Astrophys. J.* 139, 284.
Fokker, A. D. (1965). "Time and Space, Weight and Inertia." Pergamon Press, Oxford.
Fourès, Y., and Segal, I. (1955). *Trans. Amer. Math. Soc.* **78**, 385.
Gårding, L. (1947). *Ann. of Math.* **48**, 785.
Gross, L. (1964). *J. Math. Phys.* **5**, 687.
Gunn, J. E., and Oke, J. B. (1975). *Astrophys. J.* **195**, 255.
Haantjes, J. (1937). *Proc. Kon. Akad. Wetensch.* (*Amsterdam*) **40**, 700.
Hawking, S. W., and Ellis, G. F. R. (1973). "The large-scale structure of space–time." Cambridge Univ. Press, London and New York.
Hawkins, G. S. (1962). *Nature* **194**, 563.
Hetherington, N. S. (1971). Astronomical Society of the Pacific, Leaflet No. 509.
Holmberg, E. (1958). *Medd. Lunds Astron. Obs.* Ser. 2, No. 136.
Holmberg, E. (1961). *Ark. Astronom.* **2**, 559.
Holmberg, E. (1964). *Ark. Astronom.* 387.
Holmberg, E. (1969). *Ark. Astronom.* **5**, 305.
Hubble, E. P. (1929). *Proc. Nat. Acad. Sci. U.S.A.* **15**, 168.
Hubble, E. P. (1936a). *Proc. Nat. Acad. Sci. U.S.A.* **22**, 621.
Hubble, E. P. (1936b). *Astrophys. J.* **84**, 270.
Hubble, E. P., and Humason, M. L. (1931). *Astrophys. J.* **74**, 43.
Humason, M. L., Mayall, N. U., and Sandage, A. R. (1956). *Astronom. J.* **61**, 97.
Kafka, P. (1967). *Nature* **213**, 346.
Kellermann, K. I., Davis, M. M., and Pauliny-Toth, I. I. K. (1971). *Astronom. J. Lett.* **170**, L1.
Layzer, D., and Hively, R. (1973). *Astrophys. J.* **179**, 361.
Legg, T. H. (1970). *Nature* **226**, 65.
Leray, J. (1952) "Hyperbolic Partial Differential Equations." Institute for Advanced Study, Princeton, New Jersey.
Lichnerowicz, A. (1971). *General Relativity and Gravitation* **1**, 235.
Lichnerowicz, A. (1961). *Publ. Math. Inst. Haut. Etudes Sci.* No. 10.
Longair, M. S., and Rees, M. J. (1972). *Comments Astrophys. Space Phys.* **4**, 79.
Longair, M. A., and Scheuer, P. A. G. (1967). *Nature* **215**, 919.
Longair, M. S., and Scheuer, P. A. G. (1970). *Monthly Notices Roy. Astronom. Soc.* **151**, 45.
Lundmark, K. (1920). Stockholm Acad. Hand. **50**, No. 8.
Lundmark, K. (1925). *Monthly Notices Roy. Astronom. Soc.* **85**, 865.
Lynden-Bell, D. (1971). *Monthly Notices Roy. Astronom. Soc.* **155**, 119.
Lynds, R., and Petrosian, V. (1972). *Astrophys. J.* **175**, 591.
Lynds, R., and Wills, D. (1970). *Nature* **226**, 532.
Lynds, R., and Wills, D. (1972). *Astrophys. J.* **172**, 531.
Mayer, J. E. (1968). "Equilibrium Statistical Mechanics." Pergamon Press, Oxford.
Miley, G. K. (1971). *Monthly Notices Roy. Astronom. Soc.* **152**, 477.
Noonan, T. W. (1973). *Astronom. J.* **78**, 26.
Oke, J. B. (1970). *Astrophys. J. Lett.* **161**, L17.
Oort, J. H. (1974). *In* "The Formation and Dynamics of Galaxies" (Symp. No. 58). International Astronomical Union, Reidel, Dordrecht, The Netherlands.
O'Raifeartaigh, L. (1965). *Phys. Rev. Lett.* **14**, 575.
Peach, J. V. (1970). *Astronom. J.* **159**, 753.

Pearson, E. S., and Hartley, H. O. (1972). "Biometrika Tables for Statisticians," Vol. II. Cambridge Univ. Press, London and New York.
Peterson, B. A. (1970a). *Astronom. J.* **75**, 695.
Peterson, B. A. (1970b). *Astrophys. J.* **159**, 333.
Philips, T. O., and Wigner, E. P. (1968). *In* "Group Theory and Its Applications" (E. M. Loebl, ed.), Vol. I, p. 631. Academic Press, New York.
Pimenov, R. I. (1970). "Kinematic Spaces" (Seminars in Mathematics, Vol. 6). Steklov Mathematics Institute, Leningrad; translated and published by Consultants Bureau, New York.
Pooley, G. G., and Ryle, M. (1968). *Monthly Notices Roy. Astronom. Soc.* **139**, 515.
Purcell, E. M. (1969). *Astrophys. J.* **158**, 433.
Pugh, C. C. (1967). *Amer. J. Math.* **89**, 1010.
Rubin, V. C., Ford, W. K., and Rubin, J. S. 1973, *Astrophys. J.* **183**, L111.
Rees, M. J., and Sargent, W. L. W. (1972). *Comments Astrophys. Space Phys.* **4**, 7.
Robb, A. A. (1936). "Geometry of Space and Time." Cambridge Univ. Press, London and New York.
Rust, B. W. (1974). "The use of supernovae light curves for testing the expansion hypothesis and other cosmological relations." Thesis, Department of Astronomy, Univ. of Illinois, Urbana, Illinois.
Sandage, A. R. (1967). *Astrophys. J.* **150**, L9.
Sandage, A. R. (1970). *Phys. Today* **34**,
Sandage, A. R. (1972a). *Astrophys. J.* **173**, 485.
Sandage, A. R. (1972b). *Astrophys. J.* **178**, 1.
Sandage, A. R. (1972c). *Astrophys. J.* **178**, 25.
Sandage, A. (1973). *Astrophys. J.* **180**, 687.
Sandage, A., and Tammann, G. A. (1975). *Astrophys. J.* **197**, 265.
Sandage, A., Tammann, G. A., and Hardy, E. (1972). *Astrophys. J.* **172**, 253.
Sargent, W. L. W. (1971). *In* "Nuclei of Galaxies." *Vatican Conf., 1970* p. 81.
Sargent, W. L. W. (1972). *Astrophys. J.* **173**, 7.
Schmidt, M. (1968). *Astrophys. J.* **151**, 393.
Schmidt, M. (1970). *Astrophys. J.* **162**, 371.
Schmidt, M. (1972a). *Astrophys. J.* **176**, 273.
Schmidt, M. (1972b). *Astrophys. J.* **176**, 289.
Schmidt, M. (1972c). *Astrophys. J.* **176**, 303.
Scott, E. L. (1957). *Astrophys. J.* **62**, 248.
Segal, I. (1951). *Duke Math. J.* **18**, 221.
Segal, I. (1967a). *Proc. Nat. Acad. Sci. U.S.A.* **57**, 194.
Segal, I. (1967b). *J. Functional Analysis* **1**, 1.
Segal, I. (1971). *Bull. Amer. Math. Soc.*
Segal, I. (1972). *Astronom. and Astrophys.* **18**, 143.
Segal, I. E. (1975). *Proc. Nat. Acad. Sci. U.S.*, **72**, 2473.
Setti, G., and Woltjer, L. (1973a). *Astrophys. J.* **181**, L61.
Setti, G., and Woltjer, L. (1973b). *Ann. N. Y. Acad. Sci.* **223**, 8.
Shepard, G. C. (1966). *Canad. J. Math.* **18**, 1294.
Snyder, H. S. (1947). *Phys. Rev.* **71**, 38.
Stannard, D. (1973). *Nature* **246**, 295.
Takase, B. (1972). *Publ. Astronom. Sci. Japan* **24**, 295.
Tifft, W. G. (1972). *Astrophys. J.* **175**, 613.
Tits, J. L. (1957). *Forschungsinst. Math.* **1**, 98.
Tits, J. (1960). "Colloque sur la théorie de la relativité." Centre Belge Recherches Math., 107.
Van Den Bergh, S. (1970). *Nature* **225**, 503.

de Vaucouleurs, G. (1970). *Science* **167**, 1203.
de Vaucouleurs, G. (1972). *In* "External Galaxies and Quasi-Stellar Objects" (Symp. No. 44), p. 353. International Astronomical Union, Reidel, Dordrecht, The Netherlands.
de Vaucouleurs, G., and de Vaucouleurs, A. (1964). "Reference Catalogue of Bright Galaxies." Univ. of Texas Press, Austin, Texas.
Veblen, O. (1933). "Projective Relativitatstheorie."
Vorontsov-Vel'yaminov, B. A., and Ivanisevic, G. (1974). *Sov. Astron.* **18**, 174.
Wampler, E. J., Robinson, L. B., Baldwin, J. A., and Burbridge, E. M. (1973). *Nature* **243**, 336.
von Weizsäcker, C. F. (1951). *Astrophys. J.* **114**, 165.
Weyl, H., (1921). "Raum, Zeit, Materie." Springer-Verlag, Berlin.
Whitney, A. R., Shapiro, I. I., Rogers, A. E. E., Robertson, D. S., Clark, T. A., Goldstein, R. M., Marandino, G. E., and Vandenberg, N. R. (1971). *Science* **173**, 225.
Weinberg, S. (1972). "Gravitation and Cosmology." Wiley, New York.
Wigner, E. P. (1939). *Ann. of Math.* **40**, 149.
Wigner, E. P. (1950). *Proc. Nat. Acad. Sci. U.S.* **36**, 184.
Yang, C. N. (1947). *Phys. Rev.* **72**, 874.
Zeeman, E. C. (1964). *J. Math. Phys.* **5**, 490.
Zwicky, F. (1970). *Advan. Astronom. and Astrophys.* **7**, 228.

Index

A 6
B 7
C 8
D 9
E 0
F 1
G 2
H 3
I 4
J 5